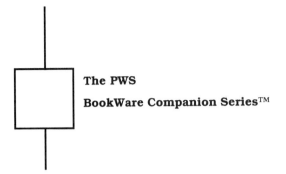

The PWS
BookWare Companion Series™

Electronics Circuit Design
Using ELECTRONICS WORKBENCH®

Muhammad H. Rashid
University of West Florida

PWS Publishing Company

I(T)P An International Thomson Publishing Company

Boston • Albany • Bonn • Cincinnati • London • Madrid • Melbourne • Mexico City
New York • Paris • San Francisco • Singapore • Tokyo • Toronto • Washington

PWS Publishing Company
20 Park Plaza, Boston, MA 02116-4324

Copyright © 1998 by PWS Publishing Company, a division of International Thomson Publishing Inc. All rights reserved. No part of this book may be reproduced, stored in a retrieval system, or transcribed in any form or by any means—electronic, mechanical, photocopying, recording, or otherwise—without the prior written permission of PWS Publishing Company.

Electronics Workbench is a registered trademark of Interactive Image Technologies Ltd.
Widows 3.1 and Windows 95 are trademarks of Microsoft Corporation.
Macintosh and Mac OS are trademarks of Apple Computer, Inc.
The BookWare Companion Series is a trademark of PWS Publishing Company.

I(T)P® International Thomson Publishing
The trademark ITP is used under license

For more information, contact:

PWS Publishing Company
20 Park Plaza
Boston, MA 02116

International Thomson Editores
Campos Eliseos 385, Piso 7
Col. Polanco
11560 Mexico C.F., Mexico

International Thomson Publishing Europe
Berkshire House 168-173
High Holborn
London WC1V 7AA
England

International Thomson Publishing GmbH
Königswinterer Strasse 418
53227 Bonn, Germany

Thomas Nelson Australia
102 Dodds Street
South Melbourne, 3205
Victoria, Australia

International Thomson Publishing Asia
221 Henderson Road
#05-10 Henderson Building
Singapore 0315

Nelson Canada
1120 Birchmount Road
Scarborough, Ontario
Canada M1K 5G4

International Thomson Publishing Japan
Hirakawacho Kyowa Building, 31
2-2-1 Hirakawacho
Chiyoda-ku, Tokyo 102
Japan

About the Cover: The BookWare Companion Series cover was created on a Macintosh Quadra 700, using Aldus FreeHand and Quark XPress.

Series Co-originators: Robert D. Strum and Tom Robbins
Editor: Bill Barter
Assistant Editor: Suzanne Jeans
Editorial Assistant: Tricia Kelly
Market Development Manager: Nathan Wilbur

Manufacturing Coordinator: Andrew Christensen
Production Editor: Pamela Rockwell
Cover Designer: Stuart Paterson, Image House, Inc
Cover Printer: Phoenix Color Corp.
Text Printer and Binder: Malloy Lithographing

Printed and bound in the United States of America.
98 99 00 01 — 10 9 8 7 6 5 4 3 2 1

Library of Congress Cataloging-in-Publication Data

Rashid, M. H.
   Electronics circuit design using Electronics workbench / Muhammad H. Rashid.
      p.   cm. — (The PWS BookWare companion series)
   ISBN 0-534-95405-7 (alk. paper)
   1. Electronic circuit design—Data processing. 2. Electronics workbench.
I. Title. II. Series.
TK7867.R39 1997
   621.3815—dc21   97-24245
                                                      CIP

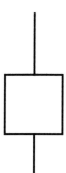

A BC Note

Students learn in a number of ways and in a variety of settings. They learn through lectures, in informal study groups, or alone at their desks or in front of a computer terminal. Wherever the location, students learn most efficiently by solving problems, with frequent feedback from an instructor, following a worked-out problem as a model. Worked-out problems have a number of positive aspects. They can capture the essence of a key concept—often better than paragraphs of explanation. They provide methods for acquiring new knowledge and for evaluating its use. They provide a taste of real-life issues and demonstrate techniques for solving real problems. Most important, they encourage active participation in learning.

We created the BookWare Companion Series because we saw an unfulfilled need for computer-based learning tools that address the computational aspects of problem solving across the curriculum. The BC series concept was also shaped by other forces: a general agreement among instructors that students learn best when they are actively inolved in their own learning, and the realization that textbooks have not kept up with or matched student learning needs. Educators and publishers are just beginning to understand that the amount of material crammed into most textbooks cannot be absorbed, let alone the knowledge to be mastered in four years of undergraduate study. Rather than attempting to teach students all the latest knowledge, colleges and universities are now striving to teach them to reason: to understand the relationships and connections between new information and existing knowledge; and to cultivate problem-solving skills, intuition, and critical thinking. The BookWare Companion Series was developed in response to this changing mission.

Specifically, the BookWare Companion Series was designed for educators who wish to integrate their curriculum with computer-based learning tools, and for students who find their current textbooks overwhelming. The former will find in the BookWare Companion Series the means by which to use powerful software tools to support their course activities, without having to customize the applications themselves. The latter will find relevant problems and examples quickly and easily and have instant electronic access to them.

We hope that the BC series will become a clearinghouse for the exchange of reliable teaching ideas and a baseline series for incorporating learning advances from emerging technologies. For example, we intend to reuse the kernel of each BC volume and add electronic scripts from other software programs as desired by customers. We are pursuing the addition of AI/Expert System technology to provide and intelligent tutoring capability for future iterations of BC volumes. We also anticipate a paperless environment in which BC content can flow freely over high-speed networks to support remote learning activities. In order for these and other goals to be realized, educators, students, software developers, network administrators, and publishers will need to communicate freely and actively with each other. We encourage you to participate in these exciting developments and become involved in the BC series today. If you have an idea for improving the effectiveness of the BC concept, an example problem, a demonstration using software or multimedia, or an opportunity to explore, contact us.

We thank you one and all for your continuing support.

The PWS Electrical Engineering Team:

BBarter@pws.com	Acquisitions Editor
SJeans@pws.com	Assistant Editor
NWilbur@pws.com	Market Development Manager
PRockwell@pws.com	Production Editor
TKelly@pws.com	Editorial Assistant

TABLE OF CONTENTS

0	INTRODUCTION	IX
1	INTRODUCTION TO ELECTRONICS WORKBENCH	1
2	DIODE CHARACTERISTICS AND APPLICATIONS	21
3	DESIGN OF A VENEER DIODE REGULATOR	33
4	DESIGN OF A DIODE RECTIFIER	43
5	DESIGN OF A COMMON EMITTER AMPLIFIER	53
6	DESIGN OF A COMMON SOURCE AMPLIFIER	63
7	DESIGN OF BUFFER AMPLIFIERS	74
8	DESIGN OF POWER AMPLIFIERS	86
9	DESIGN OF DIFFERENTIAL AMPLIFIERS	96
10	DESIGN OF OPERATIONAL AMPLIFIERS	109
11	DESIGN OF AMPLIFIERS FOR FREQUENCY RESPONSE	124
12	DESIGN OF MULTI-STAGE AMPLIFIERS FOR FREQUENCY RESPONSE	138
13	DESIGN OF ACTIVE FILTERS	148
14	DESIGN OF FEEDBACK AMPLIFIERS	158
15	DESIGN OF OSCILLATORS	172

INTRODUCTION

PURPOSES OF THIS BOOK

To provide the students:

- working knowledge of the electronics circuits simulation software, Electronics Workbench (EWB).
- techniques for modeling and simulating semiconductor devices and electronics circuits by EWB.
- skills in design of electronics circuits.
- design verification by EWB.
- analysis and performance evaluations of electronics circuits by EWB.
- pre-lab design and testing beforehand with EWB.
- laboratory skills for building and testing electronics circuits.

WHO SHOULD USE THIS BOOK?

Students in engineering and engineering technology. Also, engineers, engineering managers, or technicians who are interested in theory, analysis, design, or applications of electronic circuits and systems.

CO- OR PRE-REQUISITE

Students are expected to have a theoretical background and preparation in general electronics. A course on electronics should be a co- or pre-requisite to benefit fully from this book. It is designed to work as classroom or pre-lab design projects followed by hands-on-experience in the real lab.

COURSE DESCRIPTION

(1 credit) Diode characteristics and circuits, BJT and FET amplifiers, frequency response of single-stage and multi-stage amplifiers, feedback amplifiers, differential amplifiers, op-amp circuits, active filters, and oscillators.

RELATED TEXTBOOKS

This lab book is intended to accompany any textbook on general electronics.

DESIGN PROJECTS

There are thirteen design projects. Students are encouraged to simulate the circuits on a computer with a circuit simulation software package such as EWB. They are encouraged to build and test the projects, if possible.

BALANCE OF SCIENCE AND DESIGN CONTENT

- Engineering science: 0.25 credits or 25%.
- Engineering design: 0.75 credit or 75%.

HOW MUCH STUDY TIME IS REQUIRED?

The design projects can be integrated into either classroom lectures or laboratory courses. The material is equivalent to two typical 1-credit lab courses. That is, 3 hours of real lab per week with an additional 3 hours/week of pre-lab. Therefore, the equivalent time to be spent is 6 x 15 = 90 hours over two semesters of usually 30 weeks (or 3 terms of 30 weeks). The labs could be supplemented with an open-ended design project of 2 week duration in each semester.

HOW TO USE THIS BOOK

Read the textbook chapter(s) on the subjects listed.

Complete the design problems.

Simulate your circuit on EWB and verify your design specifications.

Test your circuits in the real laboratory, if possible.

CIRCUIT FILES

All the circuits shown in the book are stored in the enclosed diskette. The circuit files begins with prefix "FIG." For example, Figure 1-1 has the file name "FIG1_1.CA4." The extension CA4 identifies that it is drawn in EWB version 4.0. There are total of 70 circuit files.

IMPORT/EXPORT FROM SPICE®

Electronics Workbench is based on industry standard SPICE models for nonlinear analog components. You can choose from ideal or real-world models, or create your own models. You have complete control over the values and parameters of all components in a circuit. You can convert the EWB circuit to a SPICE circuit file by choosing "Export to SPICE", "Import from SPICE" from the File menu. Also, you can also export to "PCB."

ACKNOWLEDGMENTS

Thanks are due to the editorial team, Bill Barter, Monica Block, and Leslie Bondaryk at PWS Publishing for their guidance and support. I would also like to thank the reviewers for their generous time and attention:

Dr. Paul Benseker (Georgia Tech),
Dr. Thomas DeMassa (Arizona State U.),
Dr. Barry J. Farbrother (Rose-Hulman Institute of Technology),
Dr. Ward Helms (University of Washington),
Dr. James Jacob (Purdue University),
Dr. Satish M. Mahajan (Tennessee Technological University)
Dr. Thomas Plant (Oregon State University) and
Dr. Karl D. Stephan (University of Massachusetts - Amherst)

This book was prepared during my leave at King Fahd University of Petroleum & Minerals (KFUPM), Dhahran, Saudi Arabia, and I would like to thank KFUPM for giving me an academic environment conducive to scholarship and creativity.

Finally, thanks to my family for their patience while I was occupied with this and other projects.

DEDICATION

To my parents, my wife Fatema,
and
my children, Faeza, Farzana, and Hasan

ABOUT THE AUTHOR

Muhammad H. Rashid is a *Professor of Electrical Engineering* and Director of the UF/UWF Joint Program in Electrical Engineering at the University of West Florida. Dr. Rashid received a B.Sc. degree in Electrical Engineering from the Bangladesh University of Engineering and Technology, and M.Sc. and Ph.D. degrees from the University of Birmingham in the UK. Previously, he worked as a professor of electrical engineering and the chair of the Engineering Department at Purdue University at Fort Wayne. Also, he worked as a design and development engineer with Brush Electrical Machines

Ltd. (England, UK), a research engineer with Lucas Group Research Centre (England, UK), a lecturer and head of the Control Engineering Department at the Higher Institute of Electronics (Malta), a visiting assistant professor of Electrical Engineering at the University of Connecticut, an associate professor of electrical engineering at Concordia University (Montreal, Canada), and a professor of electrical engineering at Purdue University, Calumet.

Dr. Rashid is actively involved in teaching, researching, and lecturing in power electronics and has published more than 100 technical papers. He authored five Prentice-Hall books: *Power Electronics - Circuits, Devices and Applications* (1988, 2/e 1993), *SPICE for Power Electronics* (1993), *SPICE for Circuits and Electronics Using PSpice* (1990, 2/e 1995), *Electromechanical and Electrical Machinery* (1986), and *Engineering Design for Electrical Engineers* (1990). He also authored *Self-Study Guide on Fundamentals of Power Electronics* and *Selected Readings on Power Electronics*, IEEE Press, 1996. His books are adopted as textbooks all over the world. He is an internationally renowned authority on power electronics, a registered professional engineer in the Province of Ontario (Canada), a registered chartered engineer (UK), and a *fellow* of the Institution of Electrical Engineers, London. Dr. Rashid is the recipient of the *1991 Outstanding Engineer Award* from IEEE region V.

Any comments and suggestions regarding this book are welcomed and should be sent to the author.

Dr. Muhammad H. Rashid
Professor and Director
UF/UWF Joint Program in Electrical Engineering
University of West Florida
11000 University Parkway
Pensacola, FL 32514-5754
e-mail:mrashid@uwf-uf.ee.uwf.edu

1 INTRODUCTION TO ELECTRONICS WORKBENCH

1.1 LEARNING OBJECTIVES

To give an overview of Electronics Workbench and the essentials for a quick-start in building and testing electronics circuits.

After the end of this lab, you will:

- Be familiar with the Electronics Workbench, including the workspace, parts bin, and working instruments.

- Be able to build and test a circuit.

1.2 PROGRAM BASICS

Install

To install the software:

1. Place disk 1 in any available drive (A or B).
2. From Windows, enter the file manager and click the left mouse button (**CLICKL**) on drive A or B.
3. **CLICKL** on setup.exe, File, Run, OK.
4. **OK**, to select default Install Electronics Workbench (EWB).
5. **OK**, to select default C:MEWB41.
6. **Yes**, to create Electronics Workbench with icons.
7. Click the left mouse button once on the EWB icon, the window of EWB will open.

Functionality

Electronics Workbench (EWB) is an electronics lab inside a computer and it is modeled on a real electronics workbench. The large central workspace is like the breadboard, the parts bin is beside it, and the parts bin buttons and test instruments are stored on shelves along the top. You build and test circuits entirely on the workspace using the mouse and menus. Everything you need is readily at hand. Circuit behavior is simulated realistically, and the results are quickly displayed on the multimeter, oscilloscope, Bode plotter, logic analyzer, or whatever instruments you have attached to the circuit.

The general layout of the workbench is shown in Fig. 1.1 and can be divided into five areas: Window menu (top), test instruments (second from top), parts shelf (third from top), parts bin (on left column), workspace (right column), and power switch. The large central area is the workspace where you build and test a circuit. Beside the workspace is a parts bin. At the top of the display you'll find menus, test instrument icons, and the power switch for activating the circuit.

Menus   The Electronics Workbench menus are: **File**, **Edit**, **Circuit**, **Window**, and **Help**. The mouse follows an *object-action* sequence. First you select an object and then you perform an action.

Figure 1.1
General layout of Electronics Workbench

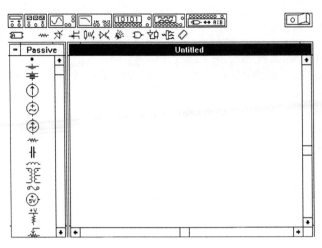

Click a menu title so it stays open. Then, click the command you want. A single click on the left mouse button *selects* an item. A double-click on the left mouse button *performs an action*. To drag a selected item, *click the left mouse button, hold down, and move the mouse*. Release the left button when the item is placed.

Parts Bin   There are multiple parts bins, which include an unlimited supply of components (*Student Edition EWB* users are limited to **25** components from the Active components parts bin), plus independent and controlled sources. The components available are determined by the parts bin button chosen.

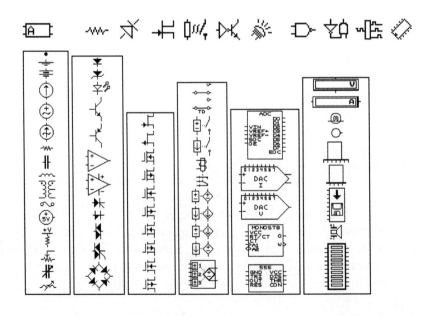

2

The parts bins include a basic parts bin (RLC and sources), active (diodes, transistors, op-amps, and thyristors), FET, control (switches and controlled sources), hybrid (DAC, ADC, 555 timer, and monostable), indicator (bulb, buzzer, LED), digital gates, combinational devices (adder, multiplexer, encoder), sequential devices (flip-flops, counter, shift register), and a customized parts bin. You can have a different customized parts bin for each schematic.

Instrument Shelf

The instrument shelf includes multimeter, function generator, oscilloscope, Bode plotter, word generator, logic analyzer, and logic converter. You can open an instrument by double-clicking the mouse on the instrument.

MULTIMETER

The simulated multimeter measures DC or AC voltage and current as well as resistance and decibel loss.

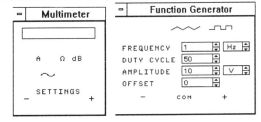

FUNCTION GENERATOR

The function generator produces sine, triangular, or square waves. You can control the signal's frequency, duty cycle, amplitude, and DC offset.

OSCILLOSCOPE

The simulated dual-channel oscilloscope behaves like the actual instrument. It supports internal or external triggering on either the positive or negative edge, and the time base is adjustable from seconds to nanoseconds. You can study hysteresis by plotting the signal magnitudes on the axes. You can also enlarge (zoom) the oscilloscope to see details and get exact readings.

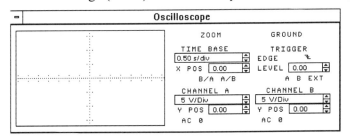

The oscilloscope displays the magnitude and frequency variations of electronic signals. It has two input terminals, channel A and channel B, so two different signals can be displayed simultaneously. You can set the oscilloscope to provide a graph of a signal's strength over time, or you can compare one signal's waveform against another's; that is, B/A or A/B.

BODE PLOTTER

Use the Bode plotter to study the frequency response of a circuit. The Bode plotter can measure either the ratio of magnitudes (voltage gain, in decibels) or phase shift (in degrees). When you specify the frequencies of interest, the plotter sweeps through them and plots the voltage gain, or phase shift, against the frequency. The plot can be displayed on either a logarithmic or linear scale. The Bode plotter generates its own frequency spectrum. The frequency of any AC sources in the circuit is ignored, but the circuit must include an AC source. Attach the Bode plotter's In and Out terminals to the nodes in the circuit at which you want to measure V_{in} and V_{out}.

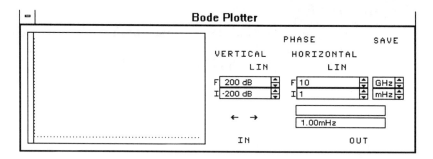

1.3 BUILDING A TEST CIRCUIT

As an example, we will draw and analyze the pulse and frequency responses of the RLC circuit shown in Fig. 1.2. The steps to build and test a circuit are:

Figure 1.2 *RLC Circuit*

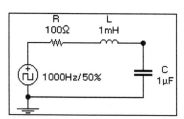

1. Drag components from the parts bin.
2. Place them on the workspace.
3. Wire the components together.
4. Level components.
5. Set component values/models.
6. Attach test instruments.
7. Activate the circuit.

Placing Components

Let us start by placing a pulse source, a resistor, an inductor, a capacitor, and a ground on the workspace. To drag an object, point to the object, press and hold the left mouse button, and move the mouse. When the object is where you want it, release the mouse button. Drag a pulse source, a resistor, an inductor, a capacitor, and a ground symbol from the parts bin, and arrange them on the workspace as shown in Fig. 1.3.

Figure 1.3 *RLC circuit components*

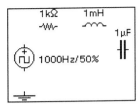

To move a component, point to it, press and hold the left mouse button, and drag the component to the new location. To remove a component, drag it back to the parts bin. Or select it and choose **Delete** from the **Edit** menu.

ROTATING COMPONENTS

Now rotate the capacitor so it can be wired neatly into the circuit. Each time you rotate a component it turns clockwise 90 degrees.

Figure 1.4 *RLC circuit components— rotated into position*

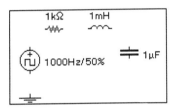

To rotate the capacitor (or other component), select it and choose **Rotate** from the **Circuit** menu.

HELP

If the **Rotate** command is dimmed, the capacitor isn't selected. Try again by pointing to it so the pointer becomes a hand, and then click the left mouse button. To deselect the selected capacitor (or other selected component), click an empty spot on the workspace with the left mouse button.

Figure 1.5 *A resistor and an inductor*

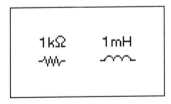

To drag or rotate two or more components at once, first select them by drawing a rectangle around the components, and then drag or rotate the rectangle. To draw a rectangle around components, point above and beside one of the components you want to select.

Press and hold the left mouse button, and drag diagonally until the rest of the components are in the rectangle that appears. To deselect one of the selected components, click it with the *right* mouse button. To deselect everything on the marked rectangle, click an empty spot on the workspace.

Wiring

Once components are placed on the workspace, you can wire them together quickly and neatly. Most components have short protruding lines called terminals that highlight when you point to them. You connect components by dragging wires from their terminals.

To wire two components, point to a terminal (a short protruding line) so it highlights. Press the left mouse button and drag so a wire appears. Stretch the wire to the other component's terminal. When it highlights, release the mouse button. The wire is automatically routed between the terminals. Wire the components as shown.

Figure 1.6
Wiring example

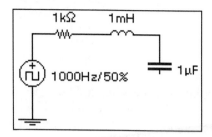

To connect two wires, use a connector. To insert a component into a wire, drag it over the wire and release the mouse button. To disconnect a wire, point to a terminal so it highlights, then drag the wire away. Release the mouse button to remove the wire altogether or to connect it to another terminal.

STRAIGHTENING A WIRE

If a wire is jagged or circuitous, there are several ways to make it straighter by moving or rotating a component to which it is attached. The method you choose depends on the cause of the problem. If components are not aligned, select one of them, and press an arrow key to align them. If a wire overlaps a component, either move the component, or drag the wire.

Rotating components often results in neater wiring. To make fine adjustments to the position of a component, select it and press an arrow key. (If you are using the arrow keys on a number pad, make sure NUM LOCK is turned off).

CONNECTING TWO WIRES

Figure 1.7
Completed RLC circuit

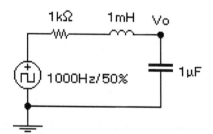

To complete this circuit, connect a wire from the capacitor's negative (lower) terminal to the wire going to the ground component.

The round dot in the parts bin is called a *connector*. Use it to connect wires to each other and to create test points in the circuit. A connector can join up to four wires, one on each side.

Drag a wire from the capacitor to the ground terminal (or where you want it to connect to the other wire), and a connector is automatically created when you release the mouse. Or, you can drag a connector from the parts bin and insert it into the wire. Then, drag a wire from the capacitor to the terminal on the connector's right side.

Drag a connector from the parts bin and insert it on the top terminal of the capacitor, and double-click on it to assign the label, Vo. Connectors are created automatically when you stretch a wire so that it touches another wire. When connectors are not needed, they disappear automatically. Permanent connectors can be created by assigning a label to them.

WIRE COLOR

To change a wire's color, select it and choose **Wire Color** from the **Circuit** menu. Or double-click the wire. Then choose a color from the box that appears.

TIP

Waveforms on the oscilloscope and logic analyzer are the same color as their probes.

Labeling

Each component in a circuit can be labeled. We will assign levels R, L, C, Vs and Vo.

Figure 1.8
Labelled circuit

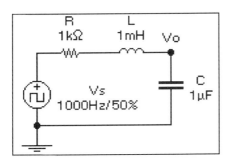

To label a component, select it by clicking the left mouse button once and choose **Label** from the **Circuit** menu. You'll see a Label box. In this case, type the label you want, then choose OK. Assign levels R, L, C, Vs, and Vo as shown.

Connectors and other components that don't have values or models can be leveled by double-clicking them. (To double-click, point to a component, and quickly press the left mouse button twice.)

HELP

If you can't see the label, the **Show labels** option is probably turned off. To turn it on, choose **Preferences** from the **Circuit** menu, and click the **Show labels** box so you see an X. Then click **Accept**.

Figure 1.9
Labelled circuit

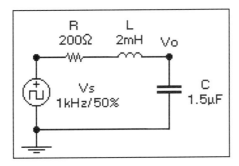

To set a component's value, select it by clicking the left mouse button once, and choose **Value** from the **Circuit** menu. You'll see a dialog box. Then type the value you want in the box that appears. If you want to change the units, click the arrows beside them. Or click inside the units box and press the up or down arrow key. Then choose **Accept**.

Component Values

Each of the components in the parts bin represents a class or type of electrical part, which you can customize to suit your needs.

> **TIP**
>
> A quick way to set a component's value is to double-click it. To set the pulse source, double-click it. Type, 10 V, 1 kHz, and 50% duty cycle, then choose **Accept**. Similarly, set $R = 200\ \Omega$, $L = 2$ mH, and $C = 1.5\ \mu F$.

HELP

If the values do not appear on the workspace, choose **Preferences** from the **Circuit** menu and turn on **Show values**.

> **TIP**
>
> If you want to use a component with the same value many times in a circuit, you can set its value in the parts bin.

Testing

To view the input and output waveforms of our circuit, we will connect an oscilloscope. There are seven test instruments stored above the workspace: multimeter, function generator, oscilloscope, Bode plotter, word generator, logic analyzer, and logic converter. Each is represented by an icon, a small picture of the instrument that you attach to test points in a circuit. The icons are stored on the instrument shelf, above the workspace. There are also two meters in the parts bin: voltmeter and ammeter.

INSTRUMENT ICONS

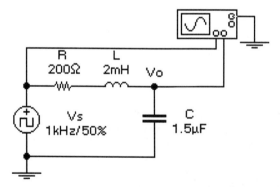

Figure 1.10 *RLC circuit with an oscilloscope attached*

To place an instrument on the workspace, point to its icon on the instrument shelf, and drag it to the workspace. To attach an instrument icon to a circuit, point to a terminal so it highlights, and drag a wire out. Attach the wire to a component.

Connect wires from the input source Vs to terminal A and from the output to terminal B of the oscilloscope. Also, connect the oscilloscope to the ground as shown. Select or move an instrument icon the same way you select or move a component.

Instrument Controls

To see an instrument, select its icon and choose **Zoom** from the **Circuit** menu. Or double-click the instrument icon. To select options, click buttons on the instrument to change values or units, and click the up or down arrows beside them.

On the function generator, oscilloscope, and Bode plotter, you can also click in the text box and press the up or down arrow key on the keyboard. For the function generator, word generator, and logic analyzer, you can select a value and type a new one.

If an instrument is hidden by another instrument, bring it forward by clicking its title bar. Double-clicking the instrument icon also brings the instrument to the front. To move an instrument, drag it by its title bar. To close an instrument, double-click its Control-menu box. To remove an instrument, drag its icon back to the instrument shelf. It will close automatically, and any wires attached to it will disconnect.

Transient vs. Steady-state Analysis

Electronics Workbench lets you analyze either the transient or steady-state response of a circuit. For the RLC circuit, we will do the transient analysis so that we can see the charging and discharging voltage of the capacitor on the oscilloscope. When a signal is first applied to a real circuit, there is a short-lived transient state before it settles down to its usual responses.

To set the analysis options, choose **Analysis Options** from the **Circuit** menu. You will see the **Analysis Options** dialog box. Select **Transient Analysis** in the **Analysis Options** dialog box if you want to analyze a circuit's behavior when the power is first turned on. The oscilloscope will display the initial, transient response of the circuit.

To see a circuit's response once steady-state has been reached, select **Steady-State Analysis** in the **Analysis Options** dialog box. If you use steady-state analysis, you may be able to shorten simulation time by specifying **Assume linear operation**. The starting conditions for the circuit are determined by its DC operating point.

Select **Transient**, select **Pause after each screen** so an **x** appears beside it. Select **Tolerance 1%**. The Tolerance setting in the **Analysis Options** dialog box controls the accuracy of the simulation results. The default setting is 1 %, which gives you the fastest simulation speed but the lowest level of accuracy. To change the level of accuracy, click **Tolerance**. Then choose another level from the menu. Tolerance values are represented exponentially; for example, 1e-5 is 0.00001.

TIP

Reducing tolerance increases the time needed to simulate a circuit. Use a 1% tolerance where possible. If you see a message saying that Electronic Workbench can't reach a solution, increase the tolerance to 1% or 10% and activate the circuit again.

Turning on the Power

To turn on the power, click the power switch in the top right corner of the display. Clicking the switch again turns off the power. Clicking the power switch activates or begins the simulation.

Saving the Circuit File

To save an unnamed circuit or make a copy of a circuit, use the **Save As** command from the **File** menu.

To use a dialog box to save a file:
1. Change directories and drives if necessary.
2. Type a file name in the text entry box. You do not need to type a file name extension.
3. If you are working in Microsoft Windows®, the circuit name must be a valid DOS file name. All EWB files are automatically given a file name extension CA4. For example, the file name FIG1_1.EWB will be saved as FIG1_1.EWB.CA4.
4. Choose **Save**. Alternately, press ENTER or ALT+S.

> ***TIP***
>
> You can also open or save a file by typing its complete pathname in the text entry box.

1.4 RLC TRANSIENT RESPONSE

Open file FIG1_1.CA4 from the EWB file menu, if you are not already in that file. Run the simulation by turning on the power. To turn on the power, click the power switch in the top right corner of the display. Clicking the switch again turns off the power.

Scope Analysis

You can see the waveforms of the input and output voltages on the oscilloscope, as in Fig. 1.11. The waveforms are shown with the setting of DC, time base 0.2 ms/div, channel A-5 V/div, channel B-5 V/div. Change the setting for a clear display.

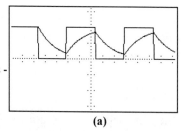
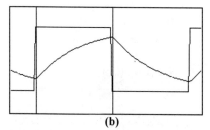

Figure 1.11 (a) Input and output waveforms (b) Input and output waveforms - zoomed

(a) (b)

Change the color of wires leading to the oscilloscope probes by double-clicking each wire and choosing a color from the box that appears. Waveforms on the oscilloscope are the same color as their probes. You can zoom the oscilloscope display as shown. You can use the cursors (one for channel A and one for channel B) to read the voltages and time axis.

Frequency Response

Now, move the pulse generator to the parts bin. Drag a sine wave generator into its place and make connections as shown. Double-click on the generator and set it to an amplitude of 10 V.

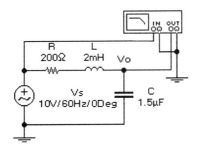

Figure 1.12 *RLC circuit with Bode plotter attached*

Move the oscilloscope to the instrument shelf. Drag the Bode plotter onto the workspace and make connections to the input and output as shown. Double-click on the Bode plotter icon to open the control panel.

1. Choose **Analysis Options** from the **Circuit** menu. Then, select **Steady-State**.
2. Save the circuit by choosing **Save As** in the **File** menu. Type a new file name in the text entry box, say FIG1_2.CW4
3. Run the simulation by clicking the power switch in the top right corner of the display.
4. You can see the frequency response on the Bode plotter. The plot is shown in Fig. 1.13 with the settings to magnitude, gain on LIN (I = 0, F = 1), and frequency on LOG (I = 1 Hz, F = 10 kHz).
5. Use the cursor on the Bode plotter to measure the break frequency f_H = 540 Hz at gain $A_{v(mid)} = V_{out}/V_{in} = 0.71$.

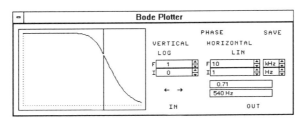

Figure 1.13 *Frequency response of RLC circuit*

1.5 CAPTURING SCHEMATICS

You can use EWB to copy circuits or output waveforms like those shown in this book to your reports.

Using the Clipboard

You can use the clipboard to transfer circuit schematics from Electronics Workbench to other applications that accept bitmap graphics. To capture the contents of the Electronics Workbench window, press ALT+PRINT SCREEN. (If the Electronics Workbench window is maximized, just press PRINT SCREEN. A bitmapped image of the Electronics Workbench window is placed on the clipboard. You can paste it directly into another application, or use the Clipboard Viewer's **Save** command to save it for later use.

Copybits Command (CTRL+I)

Use the **Copybits** command to copy all or part of the screen to the clipboard. That material may then be pasted into other applications, such as a word processor. Choose **Copybits** from the **Edit** menu; the pointer turns into a crosshair. Use the crosshair to drag a rectangle around everything you want in the picture. Release the mouse button and the area selected will be marked. Choose **Edit**, then **Copy** and the selected area will be copied to the clipboard. You can paste that material directly into another application, or use the Clipboard Viewer's **Save** command to save it for later use.

Once you have placed information on the clipboard using the **Cut** or **Copy** command, you can paste it where you want using the **Paste** command. To paste information, click the window in which you want it placed (if it is not already the active window), then choose **Paste** from the **Edit** menu. The clipboard can hold either components or text. If the clipboard does not contain information that can be pasted into the active window, the **Paste** command is dimmed. Using the **Paste** command does not affect the contents of the clipboard.

Text copied onto the clipboard from another application can be pasted into the description window. If the text was previously formatted, its formatting will be removed. Conversely, text from Electronics Workbench description windows can be pasted into other applications that accept text.

1.6 ADVANCED ANALYSIS

The EWB 5.0 version has many advanced features such as Parametric Sweep, Sensitivity Analysis, Fourier analysis, Worst-case analysis, Monte-Carlo analysis, and a graphic plot for display. If you have the earlier version of EWB, you can skip this section. The menu of EWB 5.0 is shown in Fig. 1.14.

Figure 1.14
Menu of EWB 5.0 version

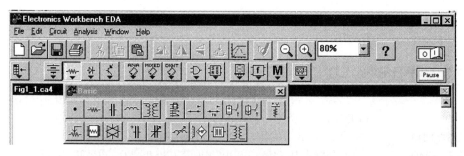

Parametric Sweep

Parameter sweep analysis allows verifying the effect of a parameter on the performance of a circuit over a range of its values. The effect is the same as simulating the circuit several times, once for each different value. The parameter values can be controlled by choosing start, end, and increment values in the Parameter Sweep dialog box.

The parametric sweep is selected from the **Analysis** menu. As an example, we will take Example 2-1 for the transient analysis of an LRC-circuit and sweep the value of R from 10 Ω to 200 Ω with an increment of 100 Ω.

1. Load your circuit such as file FIG1_1.CA4. Decide on a component and parameter to sweep and an output node for analysis. *Hint: If you point the cursor on the node, you will see the node number on the bottom display.*

Figure 1.15
Selecting parametric sweep

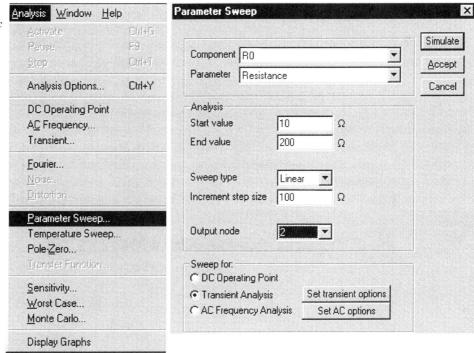

(a)   (b)

2. Choose **Parameter Sweep** from **Analysis** menu as shown in Fig. 1.15(a).
3. Choose the type of Analyis: DC Operating Point, Transient, or AC Frequency analysis.
4. Enter or change items in the dialog box as shown in Fig. 1.15(b). For example; component: R0; analysis start value: 10 Ω; analysis end value: 200 Ω; sweep type: linear; increment step size: 100 Ω; Analysis type: transient.

HELP

If you look through the list of sweep components, you will notice that the subscripts do not match with those of your circuit (i.e., R0 for R, L0 for L, and C0 for C). EWB 5.0 assigns subscripts of its own to components, such as R0, R1, R2, R3, …for resistors, L0, L1, L2, L3, … for inductors, C0, C1, C2, C3, … for capacitors, and M0, M1, M2, M3, … for meters. You can display both your symbols and the EWB assigned symbols (called reference ID) by choosing **Circuit/Schematic Options** and then enabling **Show labels** and **reference ID** as shown in Fig. 1.16(a). The schematic is shown in Fig. 1.16(b) which shows two symbols for elements such as R/R0, L/L0, and C/C0. The first one is the user defined one, whereas the second one is the EWB assigned one. Also, you can find your label and the reference ID of an element (i.e., element properties) by **DCLKR**, and then selecting **Label**, as shown in Fig. 1.17 for the properties of capacitor C/C0.

Figure 1.16
Schematics with two symbols for each element

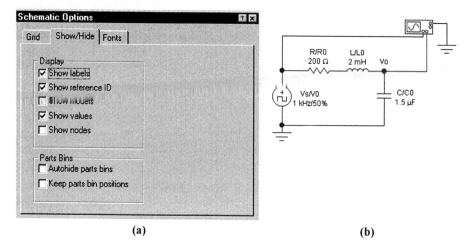

(a)   (b)

Figure 1.17
Finding element properties

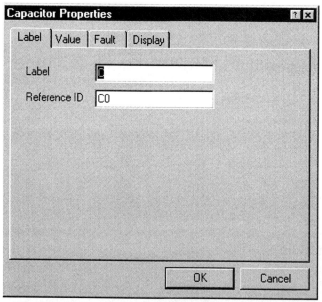

5. Click *Simulate* (press ESC to stop the analysis).

Note: If you choose transient analysis or AC frequency analysis, you can view and modify the parameters of those analyses by clicking on the **Set transient options** button or the **Set AC options** button, respectively. This will open another dialog box in which you can change the parameters for the analysis.

Parameter sweep analysis plots the appropriate curves sequentially. The **Analysis Graphs** opens automatically as shown in Fig. 1.18. As expected, for a low value of R (i.e., 10 Ω), the voltage across the capacitor exhibits decaying oscillations due to a damping factor lower than unity. However, for high values (i.e., R = 110, 200 Ω), there are no oscillations.

If the sweep type is linear, the number of curves is equal to the difference between the start and end values divided by the increment step size. If the sweep type is decade, the number of curves is equal to the number of times the start value can be multiplied by ten before reaching the end value. If the sweep type is octave, the number of curves is equal to the number of times the start value can be doubled before reaching the end value.

Figure 1.18
Plots of parametric sweep for R = 10, 110, 200 W

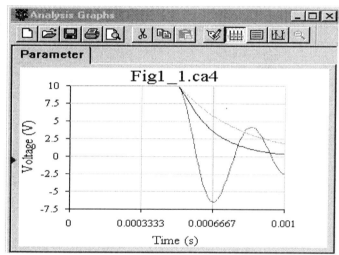

To change the graph properties such as the x-axis and y-axis, place the cursor on the graph and then click on the right mouse (**CLICKR**). Then, a dialog box will appears as shown in Fig. 1.19(a), and you select **Properties** which will allow you to choose and change the graph properties as shown in Fig. 1.19(b).

Figure 1.19
Changing graph properties

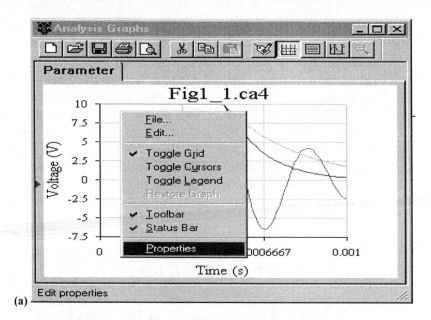

Sensitivity Analysis

Sensitivity analysis calculates the sensitivity of an output node voltage or current with respect to the parameter(s) of all components (DC sensitivity) or one component (AC sensitivity) in a circuit. Both analyses calculate the change produced in an output voltage or current by perturbing each parameter independently, and then sending the results to a table. For DC sensitivity analysis, a DC analysis is first performed to determine the DC operating point of the circuit, then the sensitivity analysis is done. For AC analysis, the AC small-signal sensitivity is calculated.

Sensitivity analysis is selected from the **Analysis menu**. As an example, we will take Example 2.2 for the AC analysis of an LRC-circuit and find the AC sensitivity of the output voltage at node 2.

1. Load your circuit such as file FIG1_2.EWB. Decide on an output voltage or current. *Hint: For an output voltage, choose nodes on either side of the circuit output. For an output current, choose a source.*
2. Choose **Sensitivity** from the **Analysis** menu.
3. Choose the type of Analysis: DC Operating Point, or AC Frequency analysis.
4. Enter or change items in the dialog box as shown in Fig. 1.20. For example, voltage at output node: 2, frequency of AC analysis, and component: R0.
5. Click **Simulate** (press **ESC** to stop the analysis).

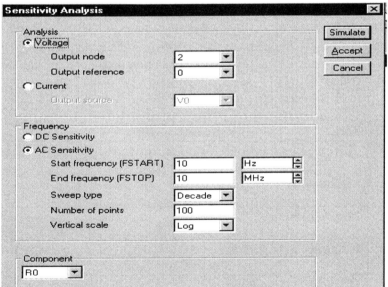

Figure 1.20
Selecting Sensitivity analysis

Sensitivity analyses produce a chart that displays the relevant parameters with their original values and their sensitivities. Sensitivity is expressed as change in output per unit change of input both in values and percentages. The **Analysis Graphs** opens automatically as shown in Fig. 1.21, which shows the AC sensitivity of the output voltage at node 2 with respect to frequency.

17

Figure 1.21
Plots of AC sensitivity analysis

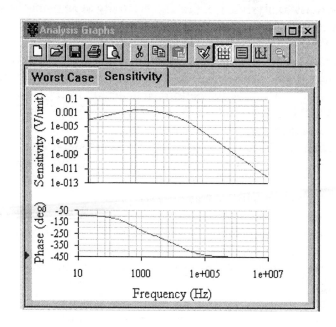

Worst-Case Analysis

Worst case analysis is a statistical analysis that gives the worst possible effects on circuit performance of variations in component parameters. The first simulation is performed with nominal values. Then, a sensitivity run (AC or DC) is performed. This allows the simulator to calculate the sensitivity of the output waveform with respect to each parameter. Once all the sensitivities have been obtained, a final run provides the worst case analysis result.

Data from the worst-case simulation is gathered by collating functions. A collating function acts as a highly selective filter by allowing only one datum to be collected per run. There are six collating functions as shown in Table 1.1.

Table 1.1

This collating function:	Captures:
Maximum voltage	the values of the Y-axis maxima.
Minimum voltage	the values of the Y-axis minima.
Frequency at maximum	the X value where the Y-axis maxima occurred.
Frequency at minimum	the X value where the Y-axis minima occurred.
Rising edge frequency	the X value the first time the Y value rises above the user-specified threshold.
Falling edge frequency	the X value the first time the Y value falls below the user-specified threshold.

Worst-case analysis is selected from the **Analysis** menu. As an example, we will take Example 2.2 for the AC analysis of an LRC-circuit and assign 500% tolerance to the parameters. Such a high tolerance is not practical, it is used as an illustration only.

1. Load your circuit such as file FIG1_2.EWB. Decide on an output node for analysis. *Hint: If you point the cursor on the node, you will see the node number on the bottom display*
2. Choose **Worst-Case** from **Analysis** menu.
3. Choose the type of Analysis: DC Operating Point, or AC Frequency analysis.
4. Enter or change items in the dialog box as shown in Fig. 1.22. For example, tolerance = 500%, collating function = Max. voltage, output node = 2, and AC analysis.
5. Click **Simulate** (press **ESC** to stop the analysis).

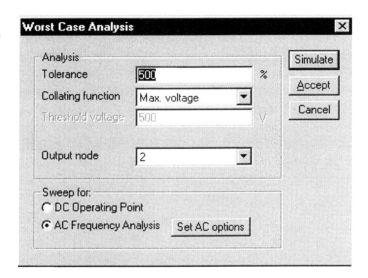

Figure 1.22
Selecting worst-case analysis

Results of the worst-case analysis are output in a chart that appears when the analysis is finished. The chart displays AC or DC response of the circuit for each run. The **Analysis Graphs** opens automatically as shown in Fig. 1.23, which shows the nominal and maximum values of the output voltage and its phase angles.

Figure 1.23
Plots of worst-case analysis with 500% tolerance

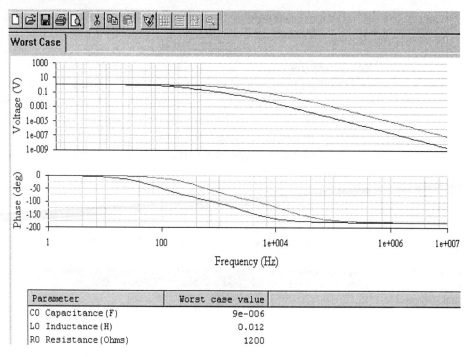

2 DIODE CHARACTERISTICS AND APPLICATIONS

2.1 LEARNING OBJECTIVES

To investigate diode characteristics and to determine the diode model parameters. We will use Electronics Workbench to plot the diode characteristic and to derive the small-signal diode parameters of a specific diode.
At the end of this lab, you will

- Be familiar with the voltage-current characteristic of a small-signal semiconductor diode.

- Be able to derive the approximate model parameters of a diode and analyze diode circuits.

2.2 THEORY

A diode is a semiconductor device that has two terminals—anode and cathode. It allows current flow in only one direction, from the anode to the cathode. If the anode voltage is higher than that of the cathode, it conducts and offers a very small resistance, typically in the order of 50Ω. If the anode voltage is lower than that of the cathode, it is said to be reverse-biased and offers a very high resistance, typically in the order of 100 k Ω. The diode current i_D is related to the anode-cathode voltage by the Shockley equation as follows:

$$i_D = I_s \left[e^{V_D/nV_T} - 1 \right] \qquad (2\text{-}1)$$

where

i_D = current through the diode from the anode to cathode, A.

v_D = diode voltage with anode positive with respect to cathode, V.

I_S = leakage (or reverse saturation) current, typically, in the range of 10^{-6} A to 10^{-15} A. At a specified temperature, the leakage current I_S will remain constant for a given diode. For small-signal (or low power) diodes, the typical value of I_S is 10^{-9} A.

n = empirical constant known as the *emission co-efficient* or *ideality factor*.

For germanium diodes, n is considered to be 1. For silicon diodes, the predicted value of n is 2, but for most practical silicon diodes, the value of n falls in the range of 1.1 to 1.8. We will assume $n = 1$ in this chapter.

V_T = a constant called the thermal voltage, given by,

$$V_T = \frac{kT}{q} \qquad (2\text{-}2)$$

where

q = electron charge, 1.6022×10^{-19} Coulombs (C).

T = absolute temperature in degrees Kelvin (°K = 273 + °C).

k = Boltzmann's constant, 1.3806×10^{-23} J per °K.

At the junction temperature of 25°C, Eq. (2-2) gives the value of V_T = 25.8 mV.

The nonlinear diode characteristic can often be represented by a piece-wise linear model as shown in Fig. 2.1(b). The diode current i_D will be very small if the diode voltage v_D is less than a specific value V_{do} known as the *threshold voltage* or the *cut-in voltage* or the *turn-on voltage* (typically 0.7 V for Si diodes). The diode will conduct fully if v_D is higher than V_{do}. Thus, the threshold voltage is the voltage at which a forward-biased diode begins to conduct. r_d is the inverse slope of the i-v characteristic and represents the current-dependent voltage. The value of r_d can be found either from the characteristic curve or the approximate relation as follows:

$$r_d = \frac{nV_T}{I_D} \qquad (2\text{-}3)$$

where I_D is the quiescent or DC operating point. Thus, the diode drop is given by

$$v_D = V_{do} + i_D r_d \qquad (2\text{-}4)$$

Figure 2.1
Diode i-v characteristics
(a) Diode symbol
(b) Piece-wise linear characteristic

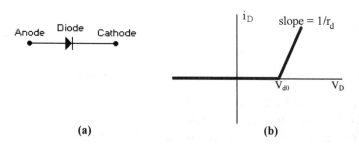

(a)　　　　　　　　　　(b)

2.3　READING ASSIGNMENT

Study chapter(s) on diodes.

2.4 PRELAB: CIRCUIT ANALYSIS

Diode Test Circuit

Figure 2.2 shows the first test circuit to plot the v-i characteristic of a diode. Resistance R_1 measures the diode current i_D and its voltage drop is $v_1 = R_1 i_D$. The input v_s is a sinusoidal voltage of 1 V peak, 100 Hz and 50% duty cycle. The input voltage varies from 0 to ±1 V in every half-cycle and it is repeated. The anode voltage is $v_2 = v_D \approx v_1 = v_D$, since v_1 is very small (i.e., mV).

The parameters of the Si diode 1N4148 are:
- Saturation current, I_S = 1.42e-8 A.
- Ohmic resistance (parasitic), r_s = 4.2 Ω.
- Zero-biased junction capacitance, Cj = 1.73 pF.
- Junction potential, Vj = 0.75 V.
- Transit time, t = 5.7 ns.
- Junction grading coefficient m = 0.333.
- Reverse breakdown voltage, V_{BR} = -74.9 V.

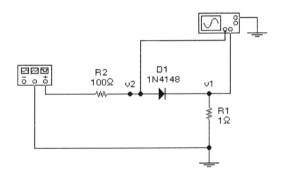

Figure 2.2 *Diode test-circuit in EWB*

Draw the theoretical v-i characteristics of the diode on the blank screen below, and also complete the calculated values in Table 2.1 in Section 2.5.

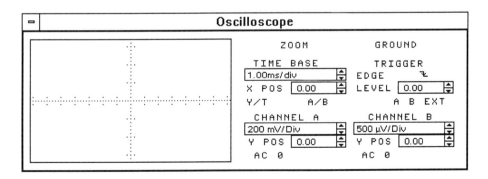

ALF-WAVE RECTIFIER

Figure 2.3 shows the second test circuit to illustrate the application of diodes in rectification. When the input voltage is greater than the threshold voltage, the diode conducts and a current flows through the load resistance R_L, and the output voltage becomes almost equal to the input voltage. On the other hand, if the input voltage is less than the threshold voltage, the diode offers a high resistance, the load current is very small, tending to be zero, and the output voltage becomes zero as well.

The particulars of the input transformer are:

- $v_s = 117$ V (peak), 60 Hz.
- Turns ratio, n = 10:1.
- Leakage inductance, $L_e = 1$ mH.
- Magnetizing inductance, $L_m = 5$ H.

Figure 2.3
Half-wave diode rectifier

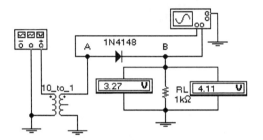

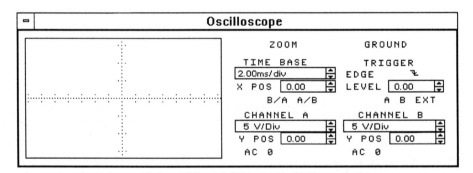

The average output voltage $V_{o(av)}$, the rms output voltage $V_{o(rms)}$, the rms ripple voltage $V_{r(rms)}$, and the ripple factor RF can be found from

$$V_{o(av)} = \frac{V_m}{\pi} \qquad (2\text{-}5)$$

$$V_{o(rms)} = \frac{V_m}{2} \qquad (2\text{-}6)$$

$$V_{r(rms)} = 1.21 V_{o(av)} \qquad (2\text{-}7)$$

$$RF = \frac{V_{r(rms)}}{V_{o(av)}} = 1.21 \tag{2-8}$$

where V_m is the peak value of the input sinusoidal voltage at the rectifier input.

Plot the output voltage (at point B) in the scope screen provided above, and also complete the calculated values in Table 2.2 in Section 2.5.

Diode Clamper

The third test circuit shown in Fig. 2.4 illustrates the application of diodes in adding a DC level to an AC input signal. However, the shape of the input signal of a clamper does not change. As soon as the input voltage v_s is switched on, the diode D_1 will be reverse-biased during the **first** positive half-cycle of the input voltage, and the output voltage will be equal to the input voltage, $v_o = v_s$. However, diode D_1 will conduct during the **first** negative half-cycle of the input voltage. The capacitor C will be charged almost instantaneously to the negative peak input voltage $(-V_m + V_{do})$, and the output voltage becomes zero, $v_o = 0$. This process is completed during the first cycle, and the circuit reaches a steady-state condition with an input voltage of $(-V_m + V_{do})$ across the capacitor C. After the first cycle, the capacitor voltage remains constant at $v_c = (-V_m + V_{do})$.

Figure 2.4
Diode clamper

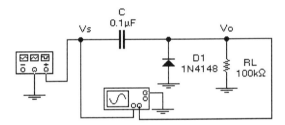

The output voltage v_o, under steady-state condition, becomes,

$$v_o = v_s + V_m - V_{do} = V_m \sin \omega t + V_m - V_{do} = V_m(\sin \omega t + 1) - v_{do}$$

for $\omega t \geq 3\pi/2$, where V_m is the peak value of the input sinusoidal voltage (5 V at 5 kHz).

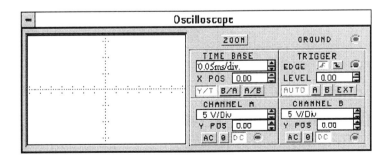

Draw the output voltage in the scope screen provided above, and also complete Table 2.3 for $V_m = 5$ V, 5 kHz, and $R_L = 100$ kΩ, 1 kΩ

2.5 EWB SIMULATION

DIODE TEST CIRCUIT

We will simulate the circuit in Fig. 2.2 in EWB. The steps to follow are:
1. Open file FIG2_2.CA4 from the EWB file menu. Run the simulation by turning on the power switch in the top right corner of the display.
2. Check that the Function Generator has settings of 1 V (sinewave) and 100 Hz. Otherwise, change the settings.
3. Check that the Oscilloscope has settings of DC; time base 1 ms/div; channel A: 200 mV/div; channel B: 500 µV/div. Otherwise, change the settings to give a clear display.
4. Zoom the oscilloscope display. Use the cursors to read for two v-i values in the linear range. One reading is already done.
5. Compare the v-i plot with your expected plot and comment:

_____

_____

_____

6. Calculate expected values of diode currents.
7. Complete Table 2.1 by calculating the small-signal resistance r_d and the threshold voltage V_{do}. If possible, breadboard and measure the practical values.

Table 2.1

	Calculated	Simulated	Practically Measured
v_{D1} (V)	0.864874 (from Eq. 2.1)	0.864874	
i_{D1} (mA)	0.51383	1.1736	
v_{D2} (V)	0.88786 (for i_{D2} = 1.2526 mA)	0.86487	
i_{D2} (mA)	1.2526	1.2526	
$r_d = \Delta v_D / \Delta i_D$ (Ω)			
$V_{do} = v_{D2} - r_d \, i_{D2}$ (V)			

If the input to the circuit in Fig. 2.2 is a DC voltage $V_{DD} = 2$ V, find the quiescent DC diode current I_D and the small-signal resistance r_d as follows:

$$I_D = \frac{V_{DD} - V_{do}}{R_1 + R_2}$$

$$r_d = \frac{25.8 \text{mV}}{I_D}$$

Does this value of r_d differ from the value obtained in Table 2.1? If yes, why?

_____

_____

_____

Half-Wave Rectifier

Simulate the circuit in Fig. 2.3 in EWB. The steps to follow are:
1. Open file FIG2_3.CA4 from the EWB file menu. Run the simulation.
2. Check that the Function Generator has settings of 117 V (sinewave) and 60 Hz. Otherwise, change the settings.
3. Check that the Oscilloscope has settings of DC; time base 2 ms/div; channel A: 5 V/div; channel B: 5 V/div. Otherwise, change the settings to give a clear display.
4. Zoom the oscilloscope display. Use the cursor to read the peak input and output voltages, and the conduction interval (time) of the diode. You can choose either DC or AC by double-clicking the voltmeter.

Table 2.2

	Calculated	Simulated	Practically Measured
$V_{o(peak)}$ (V)	11.700	11.698	
$V_{o(av)}$ (V)	3.724	3.27	
$V_{o(rms)}$ (V)	5.85	4.11	
$V_{r(rms)}$ (V)	4.51		
RF	1.21		
$t_{on(diode)}$ (ms)		8.67ms	

5. Compare the output voltage plot with your expected plot and comment:

_____

_____

6. Calculate expected values of the peak input and output voltages.
7. Complete Table 2.2 by calculating the conduction interval (time) of the diode. If possible, breadborad and measure the practical values. The conduction angle is given by

$$\theta = 2\cos^{-1}(V_{do}/V_m)$$

If the input to the circuit in Fig. 2.2 is a DC voltage $V_{DD} = 2$ V, find the quiescent DC diode current I_D and the small-signal resistance r_d as follows:

$$I_D = \frac{V_{DD} - V_{do}}{R_1 + R_2}$$

$$r_d = \frac{25.8\text{mV}}{I_D}$$

Does this value of r_d differ from the value obtained in Table 2.1? If yes, why?

_____

_____

_____

Diode Clamper

Simulate the circuit in Fig. 2.4 in EWB. The steps to follow are:
1. Open file FIG2_4.CA4 from the EWB file menu. Run the simulation.
2. Check that the Function Generator has settings of 5 V (sinewave) and 5 kHz. Otherwise, change the settings.
3. Check that the Oscilloscope has settings of DC; time base 0.05 ms/div; channel A: 5 V/div; channel B: 5 V/div. Otherwise, change the settings to give a clear display.
4. Zoom the oscilloscope display. Use the cursors to read the peak and peak-to-peak of input and output voltages.
5. Compare the output voltages with your expected plot and comment:

_____

_____

_____

6. Calculate the expected values of peak and peak-to-peak of input and output voltages.
7. Complete Table 2.3. If possible, breadborad and measure the practical values.

Table 2.3

	Calculated	Simulated	Practically Measured
Vs(peak) (V)			5v
Vs(pk-pk) (V)			10v
Vo(peak) (V) for $R_L = 100$ kΩ			9.122v
for $R_L = 1$ kΩ			6.29v
Vo(pk-pk) (V) for $R_L = 100$ kΩ			9.9891v
for $R_L = 1$ kΩ			7.2675v

8. Run the simulation for $R_L = 1$ kΩ. What is the effect of lower R_L on the output waveform? Why? Explain? How do you choose the combination of C and R_L? (Hint: $10T = C\, R_L$)

_____
_____
_____

2.6 WRITE UP/CONCLUSIONS

Write a brief report, summarizing the results of the three-step experiments and what you have learned or confirmed about diode characteristics and applications.

Comment on the relationship between theory, simulation, and practical circuit performance and the precautions that must be taken when using this learning approach.

_____
_____
_____
_____

2.7 REINFORCEMENT EXERCISES

1. A diode circuit is shown in Fig. 2.5. Plot the output voltage v_o for sinusoidal 5 V (peak), 5 kHz.

Figure 2.5
Diode circuit

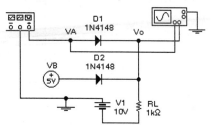

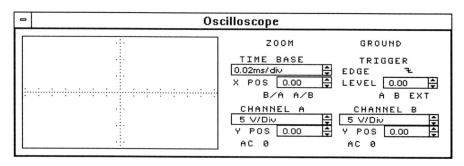

2. A diode logic circuit is shown in Fig. 2.6. Find the output voltage v_o for different values of input voltages and complete Table 2.4.

Figure 2.6
Diode logic circuit

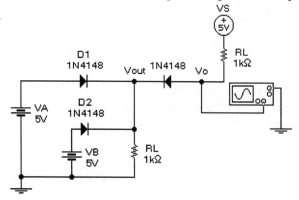

Table 2.4

V_A (V)	V_B (V)	V_{out} (V)
5	0	
0	5	
5	5	

3. A diode circuit is shown in Fig. 2.7. Plot the output voltage v_o for sinusoidal 100 mV (peak), 5 kHz and compare it with v_s. What causes the small difference, and why?

_____

_____

_____

Figure 2.7 *Diode logic circuit*

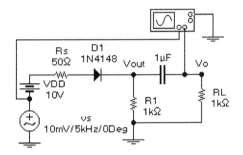

2.8 DESIGN PROBLEMS

There may be more than one solution to the following problems. Use EWB or PSpice to verify your design. Also, build and test, if possible. Determine the voltage and current ratings of active and passive components.

1. Design a clipping or limiting circuit. The input voltage is $v_s = 10 \sin(2000\pi t)$ and the output voltage should be limited to ± 4.7V. The peak current draws from the input source should be limited to 1 mA.
2. Design a clamping circuit whose input voltage is $v_s = 10 \sin(2000\pi t)$ and the output voltage should be positive in the range from 0 to 20 V. The peak current draws from the source should be limited to 1 mA.
3. Design a diode demodulator. The carrier frequency of a radio signal is fc = 250 kHz and the modulating frequency is f_m = 10 kHz. The load resistance of the demodulator is R = 10 kΩ.
4. A signal has a DC component and an ac component, such that $v_S = 10$ V + 10 mV sin (10000πt). Design the diode detector circuit shown in Fig.

2.7, which will give only the ac component within an accuracy of at least 98% at a load resistance of $R_L = 10\ k\Omega$.

2.9 DESIGN REPORT

In your report,
- Give the complete design including the ratings and values of the each component.
- Justify the use of a particular circuit topology.
- Simulate your circuit using EWB, or SPICE/PSpice to verify your design objectives including worst-case analysis with 10% tolerances for all passive components.
- Give the cost estimate. The project should be least expensive.

Suggested Report Format of Design Projects:

Table of Contents

Title Page (including your name, course number, date and year)
1. Design Objectives and Specifications
2. Design Steps (including the circuit topology)
3. Design Modifications
4. Computer Simulation and Design Verifications
5. Components and Costs
6. Flow-Chart for Design Process
7. Costs Versus Reliability and Safety Considerations
8. Conclusions

References

Appendix

3 DESIGN OF A ZENER DIODE REGULATOR

3.1 LEARNING OBJECTIVES

To design a Zener diode regulator to give a specified output voltage even though the input voltage and the load current may vary widely. We will use Electronics Workbench to verify the design and evaluate the performance of the regulator.

After the end of this lab, you will

- Be familiar with the voltage-current characteristic of a Zener diode and its small-signal model.
- Be able to analyze and design a Zener regulator to meet certain specifications.

3.2 THEORY

A Zener diode is a silicon pn-junction device that is optimized by internal design for operation in the breakdown region. It maintains an essentially constant voltage across its terminals over a specified range of reverse current values. Therefore, a minimum value of reverse current must be maintained in order to keep the diode in regulation. There is also a maximum current above which the diode will be damaged due to excessive power dissipation. In the forward direction, it exhibits the characteristic of a normal diode.

CHARACTERISTICS OF A ZENER DIODE

Let us examine the characteristics of Zener diode type 1N751, which we will use to design a voltage regulator and whose specifications are $V_z = 5.1$ V at $I_z = 20$ mA, Zener resistance $r_z = 17\ \Omega$, and $I_z(max) = 70$ mA. We will use EWB to plot the v-i characteristic and to confirm these specifications. Let us take the test circuit shown in Fig. 3.1(a). The steps to follow are:

1. Open file FIG3_1.CA4 from the EWB file menu. Run the simulation by clicking the turn-on switch.

2. Check that the Function Generator has settings of 20 V (sine wave) and 100 Hz. Otherwise, change the settings.

3. Check that the Oscilloscope has settings of DC; time base 1 ms/div; channel A: 1 V/div; channel B: 5 mV/div. Otherwise, change the settings to give a clear display.

4. Zoom the oscilloscope display. Use the cursor to read the v-i values at the knee point at which the Zener action begins. Take two other values in

the linear range. The Zener characteristic (i versus v) is shown in Fig. 3.1(b).

5. Assuming a linear relation, calculate the expected value of the voltage at the knee point.

$$V_{ZO} = V_Z - r_Z I_Z = 5.1 - 17 \times 20\text{mA} = 4.76\text{V}$$

6. Complete Table 3.1 by calculating the small-signal resistance r_z and the knee voltage V_{ZO}. V_{ZK} is the voltage and I_{ZK} is the current at the knee point..

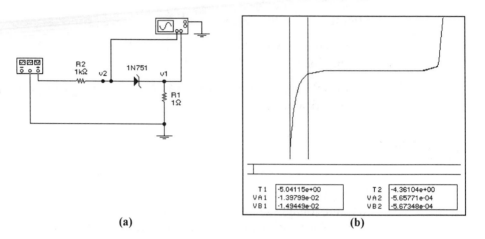

Figure 3.1
Zener diode test circuit
(a) Test circuit
(b) v-i characteristic for diode 1N751

(a)   (b)

Table 3.1

	Calculated	Simulated	Measured
V_{Z1} (V)	(for 5.86178 mA)	4.85466	
I_{Z1} (mA)	5.86178	5.86143	
V_{Z2} (V)	5.1	5.04115	
I_{Z2} (mA)	20	13.9799	
$r_z = v_z/i_z$ (Ω)	17	23	
$V_{ZO} = v_{D2} - r_z I_{Z2}$ (V)	4.76	4.72	
V_{ZK} (V)	--	4.36087	
I_{ZK} (mA)	--	0.56577	

The Zener characteristic can be represented by a piece-wise linear model. Thus, the diode voltage becomes

$$v_Z = V_{ZO} + r_z i_Z \qquad \text{in the Zener region} \qquad (3\text{-}1)$$

$$v_D = V_{do} + r_d i_D \qquad \text{in the forward direction} \qquad (3\text{-}2)$$

Note : $v_D = -v_Z$, and $i_D = -i_Z$

ZENER DIODE REGULATOR

One of the common applications of Zener diodes is in voltage regulators which can maintain a constant output voltage while both the input voltage and the load conditions may vary. This is shown in Fig. 3.2. If the input voltage is less than the Zener voltage, then the Zener diode will offer a high resistance and the output voltage will follow the potential divider rule. That is,

$$V_o = \frac{R_L}{R_L + R_S} V_{DD} \qquad \text{for } V_{DD} < V_Z \qquad (3\text{-}3)$$

If the input voltage is more than the Zener voltage, the Zener diode will offer a variable resistance such that output voltage is kept constant due to Zener action. Load resistance R_L may vary from infinity (open-circuit) to a minimum value for a specified maximum load current. The supply voltage V_{DD} may also vary from its nominal value. The value of R_S should be selected such that the Zener current must not exceed the maximum current rating when the load is disconnected and the supply voltage reaches its maximum value. Also, the Zener current must not go below the knee value when the supply voltage becomes minimum and the load current is maximum. This will ensure that the Zener operates in regulation. Ideally, the output voltage is supposed to remain constant at the Zener voltage. However, the output voltage V_o changes with the load current and the DC input voltage. It is given by

$$V_o = \frac{R_S}{R_S + r_z} V_{ZO} + \frac{r_z}{R_S + r_z} V_{DD} - I_L (R_S \| r_z) \qquad \text{for } V_{DD} > V_Z \qquad (3\text{-}4)$$

where I_L is the load current. Thus, for a specified Zener diode, R_S and V_{DD}, the output voltage will decrease with the load current. This decrease of the output voltage due to the load current is defined as the *load regulation*, which is given by

$$\text{Load Regulation (LRG)} = -\frac{V_{O(\text{No-load})} - V_{O(\text{Full-load})}}{V_{O(\text{Full-load})}} \qquad (3\text{-}5)$$

$$= -(R_S \| r_z) / R_L \qquad (3\text{-}6)$$

where the negative sign signifies decrease in the output voltage.

Figure 3.2
Zener diode regulator

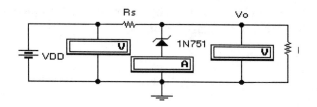

3.3 READING ASSIGNMENT

Study chapter(s) on Zener diodes.

3.4 ASSIGNMENT

Design Specifications

Design the Zener regulator shown in Fig. 3.2 to meet the following specifications:

- Output voltage, $V_O = 5.1$ V.
- DC supply voltage, $V_{DD} = 20$ V to 30 V.
- Load current, $I_L = 0$ to 40 mA.

SELECTING THE ZENER DIODE

Let us choose the Zener diode 1N751 that seems to satisfy the voltage and current specifications.

- Zener voltage, $V_Z = 5.1$ V at $I_Z = 20$ mA.
- Zener resistance, $r_Z = 17$ Ω.
- Maximum Zener current, $I_{Z(max)} = 70$ mA.
- Maximum power dissipation, $P_{D(max)} = 400$ mW.
- Zener threshold voltage, $V_{ZO} = 4.76$ V.

FINDING THE VALUE OF THE RESISTOR R_S

Under fully loaded conditions, the Zener diode will carry the minimum current, which must be adequate to ensure operation in regulation. However, under no-load conditions, the current that was to flow through the load must now flow through the Zener diode.

For designing a regulator, the minimum Zener current is usually kept at $I_{Z(min)} = I_{L(max)}/10 = 40/10 = 4$ mA. The value of R_s can be found from

$$R_S = \frac{V_{DD(min)} - V_{ZO} - r_Z I_{Z(min)}}{I_{L(max)} + I_{Z(min)}} = \frac{20 - 4.76 - 17 \times 4\text{mA}}{40\text{mA} + 4\text{mA}} \equiv 337\,\Omega$$

When $V_{DD(max)}$ is maximum, the currents are related by

$$R_S(I_{Z(max)} + I_{L(min)}) = V_{DD(max)} - V_{ZO} - r_Z I_{Z(max)}$$

which can be solved for the maximum Zener current $I_{Z(max)}$ as

$$I_{Z(max)} = \frac{V_{DD(max)} - V_{ZO} - R_S I_{L(min)}}{R_S + r_Z} = \frac{30 - 4.76 - 337 \times 0mA}{337 + 17} = 71.3mA$$

The power rating P_R of R_s is

$$P_R \approx (I_{Z(max)} + I_{L(min)})(V_{DD(max)} - V_Z) = (71.3mA + 0mA)(30 - 5.1) = 1.8W$$

The power rating P_D of the Zener diode is

$$P_D \approx I_{Z(max)} V_Z = 71.3mA \times 5.1 = 364mW$$

which is within the limit of 400 W.

FINDING THE VALUE OF THE LOAD RESISTOR R_L

From $I_{L(min)} = 0$ mA and $I_{L(max)} = 40$ mA, the maximum and minimum values of load resistances are

$$R_{L(max)} = V_Z / I_{L(min)} = 5.1 / 0mA = \infty \Omega \text{ (No - load)}$$

$$R_{L(min)} = V_Z / I_{L(max)} = 5.1 / 40mA = 127.5\Omega \text{ (Fully loaded)}$$

3.5 EWB SIMULATION

The EWB has many circuit models for many types of Zener diodes. First, you get the schematic of the Zener diode from the list of Active devices. Then, you double-click on the diode and the library of Zener diodes will open. The library has models of many commercially available diodes. Also, you can change the model parameters by choosing **Edit**.

Output Voltage We will use EWB to simulate the regulator shown in Fig. 3.1 to verify the design specifications and find its transfer characteristics. The steps to follow are:

1. Open file FIG3_2.CA4 from the EWB file menu. Run the simulation.

2. Check that the meters is set to DC. Double click the meter and then change if needed.

3. Complete Table 3.2 by calculating the percentage (%) regulation.

4. Calculate the load regulation from $-(R_S \| r_Z) \Delta I_L = -(R_S \| r_Z)/R_L$

Does this value differ from that obtained from voltage changes? Explain:

_____
_____
_____

5. Does the load regulation depend upon the supply voltage, V_{DD}? Explain:

_____
_____
_____

Table 3.2

	Calculated				Simulated				Measured			
	V_{DD}		V_{DD}		V_{DD}		V_{DD}		V_{DD}		V_{DD}	
	20	20	30	30	20	20	30	30	20	20	30	30
R_L	1M	337	1M	337	1M	337	1M	337	1M	337	1M	337
V_o					5.26 V	5.17 V	5.36 V	5.32 V				
I_L					5.26 µA	15.3 mA	5.36 µA	15.8 mA				
I_Z					43.7 mA	28.6 mA	73.1 mA	57.5 mA				
% LGR												

Transfer Characteristics

The steps to follow are:

1. Open file FIG3_3.CA4 shown in Fig. 3.3 from the EWB file menu. Run the simulation by clicking the turn-on switch.

2. Check that the Function Generators has setting of initially at 20 V (sine wave) and 100 Hz. Otherwise, change the settings.

3. Check that the Oscilloscope has settings of DC; time base of 1 ms/div; channel A: 5 V/div; channel B: 2 V/div. Otherwise, change the settings to give a clear display.

4. Zoom the oscilloscope display. Use the cursor to read V_o and V_{DD}

5. Read the critical point $V_{o(crit)}$ ($\cong$ 4.298 V) and $V_{DD(crit)}$ ($\cong$ 15.667 V) when the Zener action begins.

6. Take two readings when the Zener diode is not in regulation and find the slope, A_{non}

7. Take two readings when the Zener diode is in regulation, and find the slope, A_{zener}

8. Complete Table 3.3 by calculating A_{non} and A_{zener}

9. Ideally, the slope ($A=\Delta V_o/\Delta V_{DD}$) should be infinity when the Zener diode is in regulation. Is this value infinity (or very large)? If not, why? What causes it? Explain.

_____

_____

_____

Figure 3.3
Test circuit for transfer characteristic

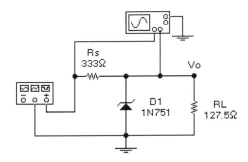

Table 3.3

	Calculated		Simulated		Measured	
	Non-Zener	Zener	Non-Zener	Zener	Non-Zener	Zener
$V_{o(crit)}$			--		--	
$V_{DD(crit)}$			--		--	
V_{o1}				--		
V_{DD1}				--		
V_{o2}			--		--	
V_{DD2}			--		--	
$A = V_o/V_{DD}$						
A_{non}						
A_{zener}						

10. Run the transient parametric sweep by varying R_s (in Fig. 3.3) from 100 Ω to 1 kΩ with an increment of 400 Ω. Set the Transient options: start time = 0, and end time = 0.02 s. Plot the transient response of the output voltage at node 1. You may need to change the scales of the x-axis and y-axis. Discuss the effects of R_s.

_____

_____

3.6 WRITE-UP/CONCLUSIONS

Write a brief report, summarizing the results of the three-step experiment and what you have learned or confirmed about Zener diode characteristics. Comment on the relationship between theory, simulation and practical circuit performance and the precautions which must be taken when using this learning approach.

_____

_____

_____

3.7 REINFORCEMENT EXERCISES

1. A Zener limiter is shown in Fig. 3.4. and it has an input voltage of sinusoidal 20 V (peak), 1 kHz. Plot the output voltage (V_o), and the transfer characteristic (V_o versus V_s) in the screens that follow.

Figure 3.4
Zener Limiter

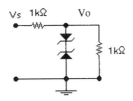

Plot output voltage

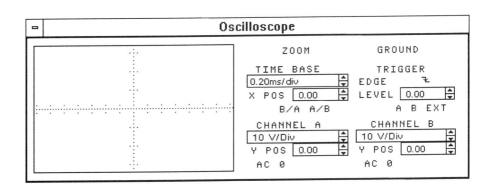

Plot transfer characteristic

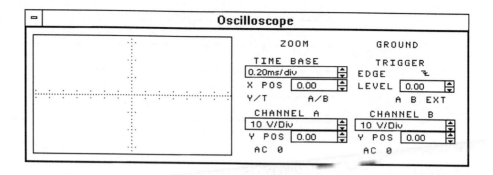

3.8 DESIGN PROBLEMS

There may be more than one solution to the following problems. Use EWB or PSPICE to verify your design. Also, build and test, if possible. Determine the voltage and current ratings of active and passive components.

1. Design a Zener regulator to give an output voltage of 6.3 V. The supply voltage V_S can vary between 12 V and 18 V, and the load current i_L changes from 5 mA to 50 mA.

2. Design a Zener regulator to give an output voltage of 4.7V ±0.1%. The supply voltage V_S can vary between 12 V and 18 V, and the load current i_L changes from 0 mA to 50 mA.

3. Design a Zener regulator to give an output voltage of ±7.8 V ± 1%. The supply voltage V_S can vary between -18 V and 18 V, and the load current i_L changes from -50 mA to 50 mA.

4 *Design Of A Diode Rectifier*

4.1 LEARNING OBJECTIVES

To design a diode rectifier in order to give a specified DC output voltage within a certain ripple voltage. We will use Electronics Workbench to verify the design and evaluate the performance of the regulator.

At the end of this lab, you will

- Be familiar with the operation of single-phase diode rectifiers, and the shapes of and relation between the input and output voltages.

- Be able to analyze and design a diode rectifier circuit to meet given specifications.

4.2 THEORY

Diodes are commonly used in rectifiers that convert an AC voltage (usually sinusoidal) to a DC voltage. The input voltage is generally the main AC supply, 120 V, 60 Hz (or 220 V, 50 Hz for British system). Electronic equipment usually requires a DC power supply in the range from 6 V to 30 V (DC). The output is much higher than this range if the rectifier is connected directly to the main supply. Normally, an input transformer is used to step down the voltage by a fixed ratio (i.e., 10:1, 25:1). Also, the transformer can provide an electrical isolation, if needed, between the supply voltage and the load. Two types of rectifiers are commonly used: center-tapped full-wave rectifiers, and bridge rectifiers.

Center-Tapped Full-Wave Rectifier The center-tapped full-wave rectifier diagram is shown in Fig. 4.1(a). This circuit uses a center-tapped transformer (hence its name), and two diodes connected to the secondary of the transformer. During the positive half-cycle of the input voltage, diode D_1 is forward-biased while diode D_2 is reverse-biased. The output voltage v_o is equal to the input voltage less the diode drop V_{do} ($\equiv 0.7$ V). During the positive half-cycle, the current flows through diode D_1 and the load resistance R_L. During the negative half-cycle of the input voltage, diode D_2 is forward-biased while diode $D1$ is reverse-biased.

Figure 4.1
Center-tapped full-wave rectifier
(a) Circuit
(b) Input and output voltages

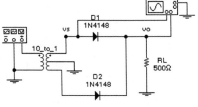
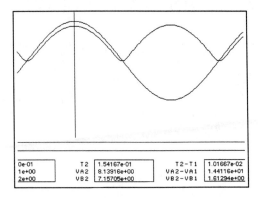

(a)   (b)

The output voltage v_o will be the inversion of the input voltage less the diode drop V_{do} ($\equiv 0.7$ V). The current flows through diode D_2 and the load resistance R_L. Thus, the circuit produces a DC (rectified) output for both the positive and negative half-cycles of the AC input voltage. The plot of the output voltage is also shown in Fig. 4.1(b) for a secondary voltage of $v_s = 16.26$ V (peak), 60 Hz. Note that the peak of the output voltage is lower than $V_m = 16.26$ V due to the diode drop. Ignoring the diode drops, the average and rms values of the output voltage are given by

$$V_{o(avg)} = \frac{2V_m}{\pi} \tag{4-1}$$

$$V_{o(rms)} = \frac{V_m}{\sqrt{2}} \tag{4-2}$$

where V_m is the peak value of the transformer secondary voltage. The rms value of the ripple voltage Vr(rms) is given by

$$V_{r(rms)} = 0.4834\, V_{o(avg)} = 0.30776 V_m \tag{4-3}$$

The *ripple factor RF* of the output voltage, which is a measure of the ripple content, can be found from

$$RF = \frac{V_{r(rms)}}{V_{o(avg)}} = \frac{0.4834 V_{o(avg)}}{V_{o(avg)}} = 0.4834 \text{ or } 48.34\% \tag{4-4}$$

Bridge Rectifier

The circuit diagram is shown in Fig. 4.2. This circuit uses four diodes and can be connected directly to the AC input supply. However, a transformer coupled input is often used. During the positive half-cycle of the input voltage, diode D_1 and D_2 are forward-biased while diodes D_3 and D_4 are reverse-biased. The output voltage v_o will be equal to the input voltage less the two diode drops $2V_{do}$ ($\equiv 2 \times 0.7$ V). The current flows through diodes D_1, D_2 and the load resistance R_L. During the negative half-cycle of the input voltage, diodes D_3 and D_4 are forward-biased while diodes D_1 and D_2 are reverse-biased. The output voltage v_o will be the negative of the input voltage less two diode drops $2V_{do}$ ($\equiv 2 \times 0.7$ V). The current flows through diodes D_3, D_4, and the load resistance R_L. Thus, the circuit produces a DC (rectified) output for both the positive and negative half-cycles of the AC input voltage. Also, the plots of the output voltage and the voltage across diode D_3 are shown in Fig. 4.2. The output is similar to that for Fig. 4.1(a). However, the peak output voltage is lower by two diode drops rather than one. For the same average output voltage, the blocking voltage of diodes in the bridge rectifier will be one-half of that for the center-tapped one.

Figure 4.2
Bridge rectifier

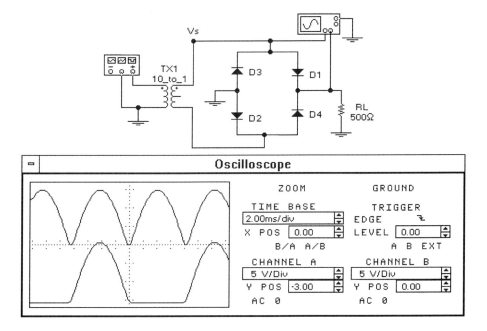

OUTPUT FILTER

The output voltage of a rectifier is not pure DC and contains harmonics. Filters are normally used to smooth out the output voltage. A capacitor C is often connected across the load to maintain a constant output voltage v_o, thereby preventing any change of voltage across the load. This is shown in Fig. 4.3.

The combination of R_1 and C_1 acts as a low-pass filter to smooth further the output voltage and also reduce the DC output voltage to the desirable value.

The peak-to-peak ripple voltage $V_{r(pk-pk)}$ can be found approximately from

$$V_{r(pk-pk)} = \frac{V_m}{2 f_S R_L C} \quad \text{for } R_1 = 0 \text{ and } C_1 \equiv 0 \quad (4-5)$$

where f_s is the frequency of the supply in Hz. The rms value of ripple voltage $V_{r(rms)}$ is given by

$$V_{r(rms)} = V_{r(pk-pk)}/2\sqrt{2} \quad (4-6)$$

The average output voltage $V_{o(avg)}$ can be found approximately from

$$V_{o(avg)} = V_m - \frac{V_m}{4 f_S R_L C} = V_m \left(1 - \frac{1}{4 f_S R_L C}\right) \quad \text{(for } R_1 = 0 \text{ and } C_1 \equiv 0\text{)} \quad (4-7)$$

Figure 4.3
Output C-filter

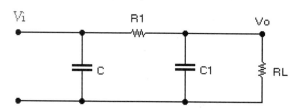

4.3 READING ASSIGNMENT

Study chapter(s) on diode rectifiers and filters.

4.4 ASSIGNMENT

Design Specifications

Design a diode rectifier to meet the following specifications:

- DC output power, $P_o = 290$ mW.
- DC output voltage, $V_o(avg) = 12$ V $\pm$ 10% ripple voltage.
- AC main supply, $V_S = 115$ V (rms), 60 Hz.

Load resistance, $R_L = V_{o(avg)}^2/P_o = 12^2/290$ mW = 497 Ω. In order to meet the specifications, one could choose either the center-tapped rectifier or the bridge one. The bridge rectifier is available commercially in a module and there is no need to build it from discrete devices. We will choose the configuration of bridge rectifier in Fig. 4.2. Thus, the complete circuit including the filter elements is shown in Fig. 4.4.

Figure 4.4
Diode rectifier with C- and low-pass filters

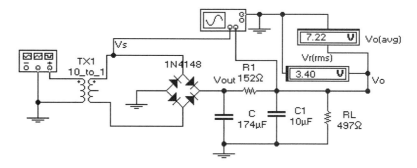

SELECTING THE INPUT TRANSFORMER

The average load current, $I_{L(avg)} = P_o/V_{o(avg)} = 290$ mW/12 V = 24 mA. The peak output voltage requirement is $V_m = \pi V_{o(avg)}/2 = \pi \times 12/2 = 18.8$ V. Since the peak supply voltage is $115 \times \sqrt{2} = 162.6$ V, we will require an input transformer with a turns ratio of $162.6/18.8 = 8.65$. Let us choose 10:1.

SELECTING THE DIODE BRIDGE

The average current rating of the bridge, $I_{L(avg)} = P_o/V_{o(avg)} = 290/12 = 24.2$ mA. The average current rating of each diode, $I_{D(avg)} = I_{L(avg)}/2 = 24.2$ mA/2 = 12.1 mA. The reverse blocking voltage of the bridge must be greater than $2 \times 162.6/10 = 24$ V. The reverse blocking voltage of each diode must be greater than $24/2 = 12$ V.

If we were to build from discrete devices, diode 1N4148 would meet the requirements. However, we will use the bridge rectifier that uses the diodes with characteristics of 1N4148.

FINDING THE VALUES OF THE OUTPUT FILTER

Follow these steps:

1. Find the peak-peak ripple voltage $V_{r(pk-pk)} = 10\%$ of $V_{o(avg)} = 0.1 \times 12 = 1.2$ V.

2. Find the value of R_1. The average output is

$$V_{o(avg)} = V_m - V_{r(pk-pk)}/2 = 16.26 - 0.6 = 15.66 \text{ V}$$

which is higher than the desired value of 12 V. Thus we need R1 whose value can be found from

$$\frac{R_L + R_1}{R_L} = \frac{15.66}{12} \text{ which gives } R_1 = 152\Omega.$$

4.5 EWB SIMULATION

The EWB has circuit models for many types of full-wave bridge rectifier diode. First, get the schematic of a bridge rectifier from the list of Active Devices. Then, you double-click on the rectifier and the library of bridge rectifier diodes will open. The library has models of many commercially available diodes. Also, you can change the model parameters by choosing **Edit**. You choose the transformer from the list of passive elements.

We will use EWB to simulate the rectifier shown in Fig. 4.4 to verify the design specifications.

The steps to follow are:

1. Open file FIG4_4.CA4 from the EWB file menu. Run the simulation.

Table 4.1

	Calculated	Simulated	Measured
$V_{o(avg)}$ for $R_1 = 152\ \Omega$	12	10.1	
$R_1 = 50\ \Omega$		11.9	
$V_{r(rms)}$ for $R_1 = 152\ \Omega$	0.424	121 mV	
$R_1 = 50\ \Omega$		212 mV	
$V_{o(rms)}$ for $R_1 = 152\ \Omega$			
$R_1 = 50\ \Omega$			
RF for $R_1 = 152\ \Omega$			
$R_1 = 50\ \Omega$			
$V_{r(pk-pk)}$ for $R_1 = 152\ \Omega$ for $R_1 = 50\ \Omega$			

2. Check that one voltmeter is set to DC to measure $V_{o(avg)}$, and other one to AC to measure $V_{r(rms)}$.

3. Check that the Function Generator has settings of sine wave, 162.6 V, and 60 Hz. Otherwise, change the settings.

4. Check that the Oscilloscope has settings of DC; time base 5 ms/div; channel A: 10 V/div, channel B: 10 V/div. Otherwise, change the settings to give a clear display.

5. Zoom the oscilloscope display. Use the cursor to read $V_{o(pk-pk)}$.

6. Complete the Table 4.1.

7. The design calculations were approximate and did not take into account the diode drops. As a result, the average output $V_{o(avg)}$ is less than the desired value of 12 V. You need to adjust the filter element. Which element would you change? Why?

_____
_____
_____

8. Run the EWB simulation with your suggested value of the element in step 7 (try changing R_1) and note the values.

$V_{o(avg)} = $ _____ $V_{r(rms)} = $ _____

4.6 WRITE-UP/CONCLUSIONS

Write a brief report, summarizing the results of the three-step experiments and what you have learned or confirmed about the full-wave rectifiers and the design of diode rectifiers.

Comment on the relationship between theory, simulation, and practical circuit performance and the precautions that must be taken when using this learning approach.

_____
_____
_____
_____
_____
_____
_____
_____

4.7 REINFORCEMENT EXERCISES

1. Take out the low-pass filter (R_1 and C_1) in Fig. 4.4. Run the simulation. Plot the output voltage below and record.

$V_{o(avg)}$ = _____ $V_{r(rms)}$ = _____

Comment on the effect of the low-pass filter on the output voltage.

_____

_____

_____

Plot for output voltage

```
 ┌─────────────────────────────────────────────────┐
 │  ■                    Oscilloscope              │
 │  ┌──────────────┐   ZOOM            GROUND      │
 │  │              │   TIME BASE       TRIGGER     │
 │  │              │   5.00ms/div      EDGE        │
 │  │              │   X POS  0.00     LEVEL  0.00 │
 │  │              │   B/A  A/B        A  B  EXT   │
 │  │              │   CHANNEL A       CHANNEL B   │
 │  │              │   10 V/Div        10 V/Div    │
 │  │              │   Y POS  0.00     Y POS  0.60 │
 │  └──────────────┘     0  DC           AC  0     │
 └─────────────────────────────────────────────────┘
```

2. A half-wave rectifier is shown in Fig. 4.5. Run the simulation. Record the following for a sinusoidal input of 162.6 V, 60 Hz.

$V_{o(avg)}$ = _____ $V_{r(rms)}$ = _____

What is the frequency of the output ripple voltage? _____ Hz.
Take out the low-pass filter (R_1 and C_1). Record.

$V_{o(avg)}$ = _____ $V_{r(rms)}$ = _____

Do R_1 and C_1 have any effect on the results? If yes, Why?

_____

_____

_____

Figure 4.5
Half-wave rectifier

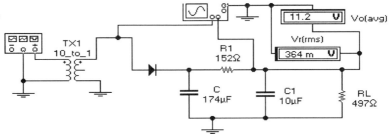

3. (Only for EWB version 5.0) Run a transient parametric sweep by varying C_1 (in Fig. 4.4) from 1 µF to 20 µF with an increment of 10 µF. Set the Transient options: start time: 0, end time: 0.02 s. Plot the transient response of the output voltage at node 2. You may need to change the scales of the x-axis and y-axis. Discuss the effects of C_1.

_____

_____

_____

4. (Only for EWB version 5.0) Run the Fourier analysis of the output voltage (at node 2) for the rectifier in Fig. 4.4. Set the dialog box as shown in Fig. 4.5: output variable at node: 2; fundamental frequency: 120 Hz (twice the supply frequency); number of harmonics: 9. Record the total harmonic distortion, THD _____ %.

Figure 4.7
Setting Fourier analysis

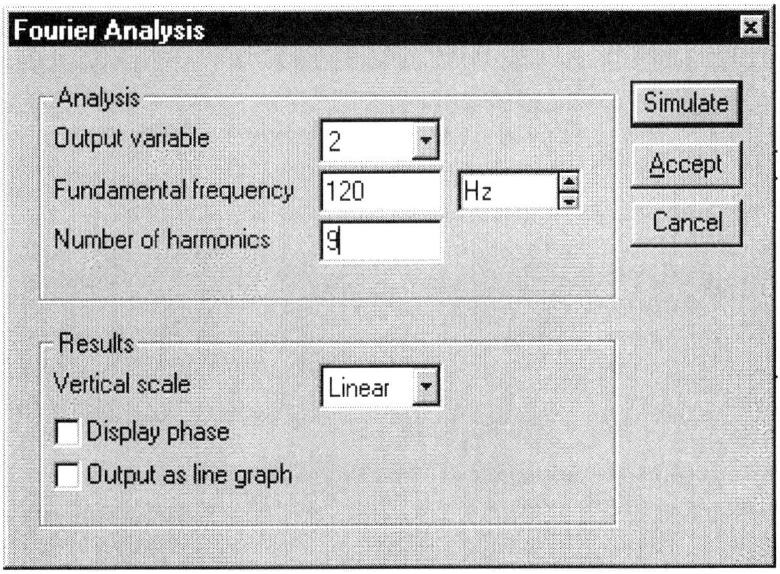

4.8 DESIGN PROBLEMS

There may be more than one solution to the following problems. Use EWB or PSPICE to verify your design. Also, build and test, if possible. Determine the voltage and current ratings of active and passive components.

1. Design a DC power supply (using single-phase half-wave rectifier) to give a DC output voltage of $V_{o(dc)} = 5$ V from an AC supply of 120 V (rms), 60 Hz. The rms ripple voltage $V_{r(rms)}$ should be limited to less than 5% of $V_{o(dc)}$. The load resistance is $R_L = 1$ kΩ.

2. Design a DC power supply (using single-phase full-wave rectifier) to give a DC output voltage of $V_{o(dc)} = 5$ V from an AC supply of 120 V (rms), 60 Hz. The rms ripple voltage $V_{r(rms)}$ should be limited to less than 5% of $V_{o(dc)}$. The load resistance is $R_L = 1$ kΩ.

3. Design a DC power supply to give a DC output voltage of 12 V from an AC supply of 120 V (rms), 60 Hz.. The load resistance is $R_L = 1$ kΩ. The ripple factor RF of the output voltage should be less than 5%. The lowest ripple frequency of the output voltage must be at least 120 Hz.

4. Design a DC power supply to give a DC output voltage of 12 V from an AC supply of 120 V (rms), 60 Hz.. The load resistance is $R_L = 1$ kΩ. The ripple factor RF of the output voltage should be less than 5%. Also, the output voltage regulation should be less than 5%. The lowest ripple frequency of the output voltage must be at least 120 Hz.

5 DESIGN OF A COMMON EMITTER AMPLIFIER

5.1 LEARNING OBJECTIVES

To design a common-emitter amplifier to give a specified voltage gain. We will use Electronics Workbench to verify the design and evaluate the performance of the amplifier.

At the end of this lab, you will

- Be familiar with the operation of a common-emitter amplifier and its small-signal equivalent circuit.
- Be able to analyze and design a common-emitter amplifier to meet certain specifications.

5.2 THEORY

A bipolar transistor can be connected in one of three configurations: common-emitter, common-base, and common collector. The common-emitter configuration gives the most voltage gain. However, in order to operate a transistor as an amplifier, it must be biased properly to establish a quiescent point, defined by the common-emitter voltage V_{CE}, the collector current I_C and the base current I_B. The input signal is then superimposed on the quiescent point by coupling capacitors, or connected directly for a direct-coupled amplifier.

Biasing of Common-Emitter Amplifier

A resistive biasing circuit which is commonly used for capacitor coupled amplifiers is shown in Fig. 5.1. Resistors R_{B1} and R_{B2} set the base voltage V_B, and hence the base current I_B which, in turn, sets the collector current I_C. Emitter resistor R_E provides the biasing stability, that is, the quiescent point becomes less sensitive to variations of the current gain β_F of the transistor. Resistor R_C sets mostly the collector emitter voltage V_{CE}. Resistors R_{B1} and R_{B2} can be replaced by a Thevenin's equivalent voltage V_{Th} and resistance R_{Th}. That is,

$$V_{Th} = \frac{R_{B2}}{R_{B1} + R_{B2}} V_{CC} \tag{5-1}$$

$$R_{Th} = \frac{R_{B1} R_{B2}}{R_{B1} + R_{B2}} \tag{5-2}$$

For a known value of V_{BE}, which is typically 0.7 V, the base current I_B can be found from,

$$I_B = \frac{V_{Th} - V_{BE}}{R_{Th} + (1+\beta_F)R_E} \qquad (5\text{-}3)$$

where β_F is the forward current gain of the transistor. The collector current I_C can be found from

$$I_C = \beta_F I_B \qquad (5\text{-}4)$$

Once the values of I_B and I_C are determined, then V_{CE} can be determined from

$$V_{CE} = V_{CC} - R_C I_C - R_E I_E = V_{CC} - R_C I_C - R_E(1+\beta_F)I_B \qquad (5\text{-}5)$$

Figure 5.1
Biasing of common-emitter amplifier

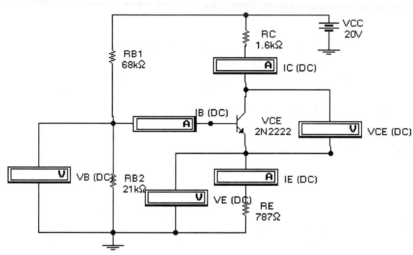

Small-Signal Amplifier

The biasing circuit is connected to a small-input signal v_s and the load resistance R_L through coupling capacitors. This is shown in Fig. 5.2. The input signal v_s is superimposed on the base voltage which causes the base current to change by a small amount. The change in the collector current is magnified due to the current gain of the transistor, that is, $\Delta i_C = \beta_F \Delta i_B$ and (in small signal quantities) $i_c = i_b$. As a result, the change in the collector voltage is magnified because $v_C = V_{CC} - R_C(I_C + \Delta i_C)$. Capacitor C_2 blocks any DC flowing to the load resistance, and the magnified change of the collector voltage (i.e., $R_C \Delta i_C$) will appear across the load. Thus, the output voltage v_o will be an amplified version of the input signal v_s, but phase shifted by 180°. The emitter capacitor C_E effectively shorts the emitter resistance R_{E2} to small-signal, and increases the voltage gain.

The small-signal base-emitter resistance r_π of the transistor is given by

$$r_\pi = \frac{\beta_F V_T}{I_C} \qquad (5\text{-}6)$$

where $V_T = 25.8$ mV. The small-signal transconductance g_m of the transistor is given by

$$g_m = \frac{\beta_F}{r_\pi} = \frac{V_T}{I_C} \qquad (5\text{-}7)$$

Figure 5.2
Common-emitter amplifier

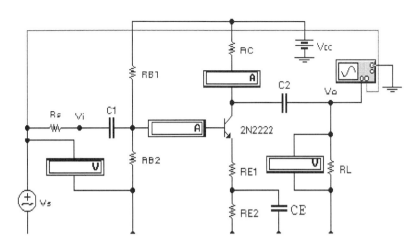

The input resistance R_i of the amplifier is given by

$$R_i = \frac{v_i}{i_S} = R_B \| \left[r_\pi + (1+\beta_F)R_E \right] \text{ where } R_B = R_{B1} \| R_{B2} \qquad (5\text{-}8)$$

The output resistance R_o of the amplifier is given by
$$R_o = R_C \qquad (5\text{-}9)$$

The open-circuit voltage gain A_{vo} (for $R_L = \infty$) is given by

$$A_{vo} = \frac{v_o}{v_s} = \frac{-\beta_F R_C}{r + (1+\beta_F) R_{E1}} \qquad (5\text{-}10)$$

The effective voltage gain A_v is given by

$$A_v = \frac{v_o}{v_s} = \frac{A_{vo} R_i R_L}{(R_i + R_s)(R_L + R_o)} = \frac{A_{vo}}{(1 + R_s/R_i)(1 + R_o/R_L)} \qquad (5\text{-}11)$$

5.3 READING ASSIGNMENT

Study chapter(s) on BJT amplifiers.

5.4 ASSIGNMENT

Design Specifications

Design the common-emitter amplifier shown in Fig. 5.1 to meet the following specifications:

- Voltage gain, $|A_v| = v_o/v_s = 100$.
- Load resistance, $R_L = 10\ k\Omega$.
- Source resistance, $R_s = 200\ \Omega$.
- DC supply, $V_{CC} = 20\ V$.
- Input signal, $v_s = 0$ to $10\ mV$.
- Use NPN-Transistor 2N2222 whose $\beta_F = 200$.
- Quiescent collector current, $I_C = 5\ mA$ (assumed).

FINDING THE VALUES OF THE BIASING RESISTORS

$\alpha_F = \beta_F/(1 + \beta_F) = 200/(1 + 200) = 0.995$.
$I_E = I_C/\alpha_F = I_C/0.995 = 5\ mA/0.995 = 5.025\ mA$.
Let us choose $V_E = V_{CC}/5 = 20/5 = 4\ V$.
$V_B = V_{Th} = V_E + 0.7 = 4.7\ V$.
$R_E = V_E/I_E = 4/5.025\ mA = 796\ \Omega$.
Let us choose $V_{CE} = 2V_{CC}/5 = 2 \times 20/5 = 8\ V$.
$R_C = (V_{CC} - V_E - V_{CE})/I_C = (20 - 4 - 8) = 8/5\ mA = 1.6\ k\Omega$.
Make $(1 + \beta_F)\ R_E$ much greater than R_{Th}. That is, as a rule of thumb,
$R_{Th} = (1 + \beta_F)\ R_E/10 = (1 + 200) \times 796/10 = 16\ k\Omega$.
$R_{B1} = R_{Th}V_{CC}/V_{Th} = 16\ k \times 20/4.7 = 68\ k\Omega$.
$R_{B2} = R_{Th}V_{CC}/(V_{CC} - V_{Th}) = 16\ k \times 20/(20 - 4.7) = 21\ k\Omega$.

DESIGNING FOR VOLTAGE GAIN

The small-signal parameters of the transistor are:
$r_\pi = \beta_f V_T/I_C = 200 \times 25.8\ mV/5\ mA = 1032\ \Omega$.
$g_m = \beta_f/r_\pi = 200/1032 = 193.8\ mA/V$.
The worst-case maximum possible no-load voltage gain that we can obtain from transistor 2N2222 operating at $I_C = 5\ mA$ is
$|A_{vo(max)}| = \beta_F R_C/r_\pi = 200 \times 1.6\ k/1032 = 310$.
The source resistance R_s will reduce the voltage. Also, the gain will be reduced when the load resistance is connected. To take into account the gain reduction, let us design for a no-load gain of 16% more (assumed). That is, for the no-load gain of $|A_{vo}| = 1.16 \times 100 = 116$.
The value of the un-bypassed emitter resistance R_{E1} in Fig. 5.2 can be found from

$$r + (1 + \beta_F)\ R_{E1} = \frac{\beta_F R_C}{|A_{vo}|} \qquad (5\text{-}12)$$

which, for $|A_{vo}| = 116$, gives $R_{E1} = 8.6\ \Omega$, and $R_{E2} = R_E - R_{E1} = 796 - 8.6 = 787.4\ \Omega$.

FINDING THE VALUES OF CAPACITORS

Capacitors C_1, C_2, and C_E set the low-cut-off frequencies, and usually have higher values. In Chapter 9, we will cover frequency response of amplifiers. Let us choose $C_1 = C_2 = C_E = 10\ \mu F$ so that they are effectively shorted at the frequency of the input signal.

5.5   EWB SIMULATION

The EWB has circuit models for many types of both pnp and npn transistors. First, you get the schematic of the transistor from the list of Active devices. Then, you double-click on the transistor and the library of npn (or pnp) transistors will open. The library has models of many commercially available transistors. Also, you can change the model parameters by choosing **Edit.**

We will use EWB to simulate both the biasing circuit and the amplifier to verify the design specifications.

Biasing Circuit

The steps to follow are:

1. Open file FIG5_1.CA4 from the EWB file menu.

2. Check that the ammeters are set to DC. Double-click the meter and then change settings, if needed.

3. Run the simulation. Complete Table 5-1 for $\beta_F = 200$.

4. You double-click the transistor 2N2222., choose **Edit,** and then change the value of $\beta_F = 300$. Repeat the simulation for $\beta_F = 300$.

5. Make R_E very small, say $0.01\ \Omega$, and run the simulation for $\beta_F = 200, 300$.

    (12.4 mA)   (12.4 mA)

$I_C\ (\beta_F = 200)$ _____ $I_C\ (\beta_F = 300)$ _____ $\%\Delta I_C$ _____

6. Comment on the effect of R_E on $\%\Delta I_C$ due to any change in β_F.

_____

_____

_____

Table 5.1

	Calculated	Simulated	Practically Measured
V_C for $\beta_F = 200$ for $\beta_F = 300$	12 V	12.51 V 12.28 V	
V_E for $\beta_F = 200$ for $\beta_F = 300$	4 V	3.70 V 3.81 V	
V_{CE} for $\beta_F = 200$ for $\beta_F = 300$	8 V	8.81 V 8.47 V	
I_B (μA) for $\beta_F = 200$ for $\beta_F = 300$	25 16.7	21.9 15.2	
I_C (mA) for $\beta_F = 200$ for $\beta_F = 300$	5	4.68 4.83	
%ΔI_C			
%ΔV_{CE}			

7. Run the DC sensitivity analysis of the collector voltage of the transistor (in Fig. 5.1) with respect to resistors. Discuss the results.

_____

_____

_____

Small-Signal Amplification

The steps to follow are:
1. Open file FIG5_2.CA4 from the EWB file menu. Run the simulation. If the simulation never reaches the steady-state condition, make it stop.by clicking the stop button.
2. Check that the input signal is 10 mV, 10 kHz. Otherwise, change the settings.
3. Check that the Oscilloscope has settings of AC, time base 0.01 ms/div, channel A: 1 V/div, channel B: 10 mV/div. Otherwise, change the settings to give a clear display
4. .Zoom the oscilloscope display. Use the cursor to read the v_o and v_s. Find the voltage gain.

        (1.158 V)   (14.14 mV)   (81.90)

$V_{o(peak)}$ _____ , $V_{s(peak)}$ _____ $A_v = V_{o(peak)}/V_{s(peak)}$ _____

5. Complete Table 5.2 by recording v_o, v_i and i_s.

6. To find the output resistance R_o, run the simulation for $R_L = 10\ M\Omega$, $10\ k\Omega$.

$$R_o = R_L(=10k\Omega)\left[\frac{v_o(\text{for }R_L=10M\Omega)}{v_o(\text{for }R_L=10k\Omega)}-1\right] \quad (5\text{-}13)$$

$R_o = 10\ k\Omega \times (96.5\ mV/835\ mV - 1) = 1557\ \Omega$.

The design calculations were approximate and did not take into account the gain drop due to R_s. As a result, the voltage gain A_v is less than the desired value of 100. Which resistance would you adjust to get the desired gain? Why?

_____

_____

7. Run the EWB simulation with the value of your suggested element in step 6 (try adjusting R_{E1} and R_{E2}).

v_o _____ v_s _____ $A_v = v_o/v_s$ _____

You reduce R_{E1} to zero, and find that the voltage gain is still less than the desired value. What should you do? How would you meet the gain specifications? Comment.

_____

_____

_____

Table 5.2

	Calculated	Simulated	Practically Measured
v_o (mV)	--	835	
v_i (mV)	--	9.28	
i_s (μA)	--	3.62	
Av = vo/vs (V/V)	100	89.98	
$R_i = v_i/i_s$ (kΩ)	2.41	2.56	
R_o (kΩ)	1.6	1.56	

8. Run the transient sweep by varying R_{E1} (in Fig. 5.2) from 0.001 Ω to 50 Ω with an increment of 25 Ω. Set the Transient options: start time: 0; end time: 0.0001 s. Plot the transient response of the output voltage at

the positive terminal of the load resistance R_L. You may need to change the scales of the x-axis and y-axis. Discuss the effects of R_{E1}.

5.6 WRITE-UP/CONCLUSIONS

Write a brief report, summarizing the results of the design experiment and what you have learned or confirmed about the common-emitter amplifier.

Comment on the relationship between theory, simulation and practical circuit performance and the precautions that must be taken when using this learning approach.

5.7 REINFORCEMENT EXERCISES

A common-emitter amplifier with an active load is shown in Fig. 5.3. Open file FIG5_3.CA4 from the EWB file menu. Run the simulation.

Table 5.3

V_s (mV)	720	725	730	735	740	745
V_o (V)	19.9V	18.4V	14.8V	10.9V	5.80V	1.04V

Figure 5.3
Common-emitter amplifier with an active load

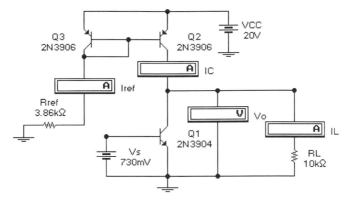

Plot for transfer characteristic (V_o versus the input voltage V_s

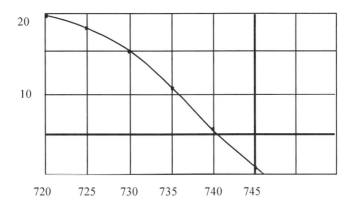

1. Record the input voltage v_s and output voltage v_o. Complete Table 5.3. From the data, plot the output voltage vo versus the input voltage vs, and calculate the small-signal voltage gain, Av = ΔVo/ΔVs. (5.8 - 1.04)/5 mV = 952.
2. Record the DC biasing current I_C and the DC reference current I_{ref}.
3. Complete Table 5.4 (for V_s = 745 mV) by calculating the DC biasing current I_C, the DC reference current I_{ref}, and the small-signal voltage gain A_v. The small-signal voltage gain can be found from

$$A_V = -g_m \left(r_{01} \| r_{02} \| R_L \right) = -1296 \tag{5-14}$$

where

$g_m = I_C/V_T$ is the transconductance of the transistor Q_1 (193.9 mA/V).
For I_C = 3.64 mA, g_m = 3.64 mA/25.8 mV = 141.1 mA/V.
$r_{01} = V_{A1}/I_C$ is the output resistance of transistor Q_1 (40.8 kΩ).
$r_{02} = V_{A2}/I_C$ is the output resistance of transistor Q_2 (40 kΩ).
V_{A1} = 204 V for the npn-transistor 2N3904, and V_{A2} = 200 V for the pnp-transistor 2N3906.

Table 5.4

	Calculated	Simulated	Ptractically Measured
A_v	1296	1340 (for V_s = 745 mV V_o = 1.04V) V_s = 742 mV V_o = 4.18V)	
I_C (mA)	5	5.09	
I_{ref} (mA)	5	5.0mA	

5.8 DESIGN PROBLEMS

There may be more than one solution to the following problems. Use EWB or PSPICE to verify your design. Also, build and test, if possible.

1. An amplifier is needed to amplify the output of a transducer which produces a voltage signal of v_s = 5 mV with an internal resistance of R_s = 2.5 kΩ. The load resistance can vary from R_L = 2 kΩ to 10 kΩ. The desired output voltage requirement is v_o = 2.5 V. The amplifier must not draw more than 1μA from the transducer. The output voltage variation when the load is disconnected should be less than 0.5%. Determine the design specifications of the amplifier (input resistance, output resistance, and gain).

2. Design a BJT-amplifier, shown in Fig. 5.2, to give a mid-frequency voltage gain of $|A_v| = |v_L/v_s|$ = 50 at a load resistance of R_L = 10 kΩ. Assume a source resistance of R_s = 500 Ω, and V_{CC} = 15 V.

3. Design a BJT amplifier, shown in Fig. 5.3, with an active current source to give a mid-frequency voltage gain of $|A_v| = v_L/v_s \geq 500$ at a load resistance of R_L = 10 kΩ. Assume a source resistance of R_s = 500 Ω, and V_{CC} = 15 V.

4. Design a BJT-amplifier, shown in Fig. 5.2, to give an input resistance of $R_{in} = v_s/i_s \geq$ = 25 kΩ. The load resistance is R_L = 10 kΩ. Assume a source resistance of R_s = 500 Ω, and V_{CC} = 12 V.

6 DESIGN OF A COMMON SOURCE AMPLIFIER

6.1 LEARNING OBJECTIVES

To design a common-source (JFET) amplifier to give a specified voltage gain. We will use Electronics Workbench to verify the design and evaluate the performance of the amplifier.

At the end of this lab, you will
- Be familiar with the operation of a common-source amplifier and its small-signal equivalent circuit.
- Be able to analyze and design a common-source amplifier to meet certain specifications.

6.2 THEORY

Like a BJT, a field-effect transistor (FET) can be connected in one of three configurations: common-source, common-gate, and common drain. The common-source configuration gives the most voltage gain, but the voltage gain is much lower than that of a comparable BJT amplifier for the same biasing current. Similar to the common-emitter (BJT) amplifier, the common-source amplifier must be biased properly to establish a quiescent point, defined by the drain-source voltage V_{DS}, the drain current I_D, and the gate-source voltage V_{GS}. The input signal is then superimposed on the quiescent gate voltage through coupling capacitors, or connected directly for direct-coupled amplifiers.

Biasing of Common-Source Amplifiers

The resistive biasing circuit used for the common-emitter amplifier can also be used for a common-source FET amplifier. This is shown in Fig. 6.1. Resistors R_{G1} and R_{G2} set the gate voltage V_G, and hence the gate-source voltage V_{GS} which, in turn, sets the drain current I_D. The source resistor R_{sr} adjusts the gate-source voltage V_{GS}. Resistor R_D sets mostly the drain-source voltage V_{DS}. Since the gate current of the JFET, J_1, is very small (tending to be zero) the gate voltage follows the potential divider rule. That is,

$$V_G = \frac{R_{G2}}{R_{G1} + R_{G2}} V_{DD} \tag{6-1}$$

The gate-source voltage V_{GS} is given by

$$V_{GS} = V_G - R_{sr} I_D \tag{6-2}$$

The gate-source voltage V_{GS} and the drain current I_D are related by

$$I_D = I_{DSS}\left(1 - \frac{V_{GS}}{V_P}\right)^2 \tag{6-3}$$

$$= K_p\left(V_{GS} - V_{to}\right)^2 \tag{6-4}$$

where I_{DSS} is the drain current for $V_{GS} = 0$ V, V_p is the pinch-off voltage of a JFET, $V_{to} = V_p$ for a JFET, and K_p is the FET constant, equal to I_{DSS}/V_p^2 for a JFET.

Once the value of I_D is determined, then V_{DS} can be determined from.

$$V_{DS} = V_{DD} - R_D I_D - R_{sr} I_D = V_{DD} - (R_D + R_{sr})I_D \tag{6-5}$$

Figure 6.1
Biasing of common-source amplifier

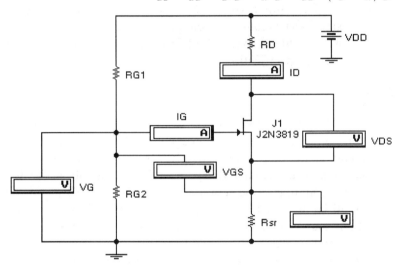

Small-Signal Amplifier

The biasing circuit is connected to a small-input signal v_s and the load resistance R_L through coupling capacitors. This is shown in Fig. 6.2. The input signal v_s is superimposed on the gate voltage which causes the gate-source voltage to change by a small amount. The change in the drain current is magnified due to the transconductance gain of the transistor, that is, $\Delta i_D = g_m \Delta v_{GS}$ and $i_d = g_m v_{gs}$. As a result, the change in the drain voltage is magnified, because $v_D = V_{DD} - R_D (I_D + \Delta i_D)$. Capacitor C_2 blocks any DC flowing to the load resistance and a magnified change of the drain voltage will appear across the load. Thus, the output voltage v_o will be an amplified version of the input signal v_s, but phase shifted by 180°. The source capacitor C_S effectively shorts the source resistance R_{sr2} to small-signals, and increases the voltage gain.

The small-signal transconductance g_m of the transistor is

$$g_m = 2K_p\left(V_{GS} - V_{to}\right) \tag{6-6}$$

Figure 6.2
Common-source amplifier

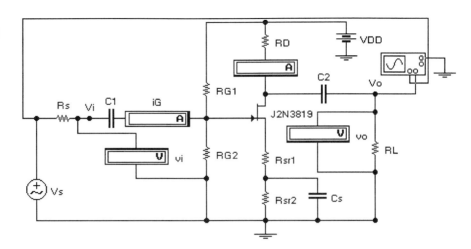

The input resistance R_i of the amplifier is given by

$$R_i = \frac{v_i}{i_s} = R_G = R_{G1} \| R_{G2} \tag{6-7}$$

The output resistance R_o of the amplifier is given by

$$R_O \equiv R_D \tag{6-8}$$

The open-circuit voltage gain A_{vo} (for $R_L = \infty$) gain is given by

$$A_{vo} = \frac{v_o}{v_g} = \frac{-g_m R_D}{1 + g_m R_{sr1}} \tag{6-9}$$

which indicates that the source resistance R_{sr1} reduces the open-circuit voltage gain A_{vo}.

The effective voltage gain A_v is given by

$$A_v = \frac{v_o}{v_s} = \frac{A_{vo} R_i R_L}{(R_i + R_s)(R_L + R_o)} = \frac{A_{vo}}{(1 + R_s/R_i)(1 + R_o/R_L)} \tag{6-10}$$

6.3 READING ASSIGNMENT

Study chapter(s) on FET amplifiers.

6.4 ASSIGNMENT

Design Specifications

Design the common-source amplifier shown in Fig. 6.2 to meet the following specifications:

- Voltage gain, $|A_v|$ = vo/vs = 10.
- Load resistance, R_L = 10 kΩ.
- Source resistance, R_s = 200 Ω.
- Input signal voltage, v_s = 0 to 100 mV.

- DC supply $V_{DD} = 20$ V.
- Use JFET transistor J2N3919 whose $I_{DSS} = 11$ mA, $V_p = -2.5$ V, and K_p (transconductance) = 1.76 mA/V².
- Quiescent drain current, $I_D = 5$ mA (assumed).

FINDING THE VALUES OF THE BIASING RESISTORS

The desired gate-source voltage can be found from

$$I_D = K_p (V_{GS} - V_{to})^2$$

which, for $V_{to} = V_p = -2.5$ V, $K_p = 1.76$ mA/V² and $I_D = 5$ mA, gives $V_{GS} = -0.8145$ V.

For a large voltage gain, the drain resistance R_D should be large. However, this will reduce the available drain-source voltage V_{DS}. For operating the JFET in the saturation region,

$$V_{DS} > (V_{GS} - V_P) = (-0.8145 + 2.5) = 1.6855 \text{ V}$$

However, we must ensure that the drain-voltage swing will not cause the JFET to move from the saturation region. For $v_s = 100$ mV and $|A_v| = 10$, this swing will be 100 mV x 10 = 1.0 V. Thus, the minimum value of $V_{DS(min)} = 1.6855 + 1.0 = 2.6855$ V. Let us choose

- $V_{DS} = 3.0$ V (assumed)
- $V_{SR} = 4$ V (assumed)

Thus, $V_G = V_{SR} + V_{GS} = 4 - 0.8145 = 3.1855$ V. From Eq. (6-1), we can find R_{G1} and R_{G2} as follows

Let $R_{G1} = 1.7$ MΩ which gives $R_{G2} = 9$ MΩ.
Thus,

$$V_G = 20 \times 1.7 \text{ M}\Omega / (1.7 \text{ M}\Omega + 9 \text{ M}\Omega) = 3.17757 \text{ V}$$

$$V_{SR} = V_G - V_{GS} = 3.17757 + 0.8145 = 3.99207 \text{ V}$$

which gives

$$R_{sr} = V_{SR} / I_D = 3.99207/5 \text{ mA} = 798.4 \, \Omega$$

$$R_D = (V_{DD} - V_{DS})/I_D - R_{sr} = (20 - 3.0)/5 \text{ mA} - 798.4 = 2.6 \text{ k}\Omega$$

DESIGNING FOR VOLTAGE GAIN

The small-signal transconductance g_m becomes

$g_m = 2 K_p (V_{GS} - V_{to}) = 2 \times 1.76 \times (-0.8145 + 2.5) = 5.93296$ mA/V.

The worst-case maximum possible voltage gain that we can obtain from the amplifier is

$$|A_{v(max)}| = g_m(R_D\|R_L) = 5.93296 \text{ mA/V} \times (2.6\text{ k}\|10\text{ k}) = 12.24$$

The value of un-bypassed source resistance R_{sr1} in Fig. 6.2 can be found from

$$1 + g_m R_{sr1} = \frac{g_m(R_D\|R_L)}{|A_v|} \tag{6-11}$$

which, for $|A_v| = 10$, gives $R_{sr1} = 37.8\ \Omega$ and $R_{sr2} = R_{sr} - R_{sr1} = 798.4 - 37.8 = 760\ \Omega$.

FINDING THE VALUES OF CAPACITORS

Capacitors C_1, C_2, and C_S set the low-cut-off frequencies and usually have higher values. In Chapter 9, we will cover frequency response of amplifiers. Let us choose $C_1 = C_2 = C_S = 10\ \mu F$ so that they are effectively short-circuited at the frequency of the input signal.

6.5 EWB SIMULATION

The EWB has circuit models for many types of both n-channel and p-channel FETs. First, you get the schematic of an FET from the list of FET devices. Then, you double-click on the JFET and the library of FETs will open. The library has models of many commercially available FETs. Also, you can change the model parameters by choosing **Edit**.

We will use EWB to simulate both the biasing circuit and the amplifier to vary the design specifications.

Biasing Circuit

The steps to follow are:

1. Open file FIG6_1.CA4 from the EWB file menu.
2. Check that the meters are set to DC. Double-click the meter and then change, if needed.
3. Run the simulation. Complete Table 6.1.

Table 6.1

	Calculated	Simulated	Practically Measured
V_{DS} (V)	3	2.32V	
V_{SR} (V)	3.992	4.05V	
V_D (V)	6.992	6.37V	
V_G (V)	3.17757	3.24V	
V_{GS} (V)	-0.8145	-0.814	
I_D (mA)	5	5.05	

4. Run the DC sensitivity analysis of the drain voltage of the transistor (in Fig. 6.1) with respect to resistors (with ±10% tolerances). Discuss the results.

_____
_____
_____

Small-Signal Amplification

The steps to follow are:
1. Open file FIG6_2.CA4 from the EWB file menu. Run the simulation.
2. Check that the input signal is 100 mV, 10 kHz. Otherwise, change the settings.
3. Check that the Oscilloscope has settings of AC (if it comes up DC, then chnage to AC); time base: 0.01 ms/div; channel A: 100 mV/div; channel B: 1 V/div. Otherwise, change the settings to give a clear display.
4. Zoom the oscilloscope display. Use the cursor to read the v_o and v_s. Find the voltage gain

(1.266 V)　　　(0.141 V)　　　　　　　　　(8.98)

$V_{o(peak)}$ _____　$V_{s(peak)}$ _____　$A_v = V_{o(peak)}/V_{s(peak)}$ _____

5. Complete Table 6.2 by recording v_o, v_i and i_s.
6. To find the output resistance R_o, run the simulation for R_L = 10 MΩ, 10 kΩ.

$$R_o = R_L(=10\text{ k}\Omega)\left[\frac{V_o(\text{for } R_L = 10 \text{ M}\Omega)}{V_o(\text{for } R_L = 10 \text{ k}\Omega)} - 1\right] \qquad (6\text{-}12)$$

R_{o}, = 10 kΩ. x (1.07 V/900 mV - 1) = 1.89 Ω.

7. The design calculations were approximate and did not take into account the gain drop due to R_s. As a result, the voltage gain A_v is expected to be less than the desired value of 10. Would R_s affect the gain significantly? If not, why? Explain.

_____
_____
_____

8. If the voltage gain is more or less than the desired value of 10, which resistance would you adjust to get the desired gain? Why?

_____

_____

_____

9. Run the EWB simulation with the value of your suggested element in step 8 (try adjusting R_{sr1} and R_{sr2})

v_o _____ v_s _____ $A_{(v = vo)}/v_s$ _____

10. You reduce R_{sr1} to zero, and find that voltage gain is still less than the desired value. What should you do? How would you meet the gain specifications? Comment.

_____

_____

_____

11. Change the input voltage to $v_s = 200$ mV by double-clicking the source v_s. Observe the output voltage on the oscilloscope. Is the output distorted or clipped? Why? What would you do to avoid distortion? Explain.

_____

_____

_____

Table 6.2

	Calculated	Simulated	Practically Measured
v_o (V)	--	1.266V	
v_i (mV)	--	0.141V	
i_s (mA)	--	5.02	
$A_v = v_o/v_s$ (V/V)	10	8.98	
$R_i = v_i/i_g$ (MΩ)	1.43	100 mV/0.15 µA = 0.67 MΩ	
R_o (kΩ)	2.6	1.889	

12. Run the transient parametric sweep by varying R_{sr1} (in Fig. 6.2) from 0 Ω to 100 Ω with an increment of 50 Ω. Set the Transient options: start time = 0, and end time = 0.0001 s. Plot the transient response of the output voltage at node 1. You may need to change the scales of the x-axis and y-axis. Discuss the effects of R_{sr1}.

_____

_____

_____

6.6 WRITE-UP/CONCLUSIONS

Write a brief report, summarizing the results of the design experiment and what you have learned or confirmed about the common-source amplifier.

Comment on the relationship between theory, simulation and practical circuit performance and the precautions that must be taken when using this learning approach.

_____

_____

_____

_____

_____

_____

6.7 REINFORCEMENT EXERCISES

1. If the biasing resistance R_{G1} in Fig. 6.2 is removed, the circuit is known as the self-biasing circuit. This is shown in Fig. 6.3.

    A. Open file FIG6_3.CA4 from the EWB file menu.
    B. Run the simulation and complete the Table 6.3. Check that meters for v_o and v_s are set to AC and others to DC values.

Figure 6.3
A self-biasing common-source JFET amplifier

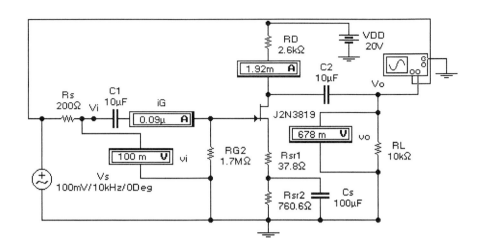

Table 6.3

	Calculated	Simulated	Practically Measured
I_D	--	1.92 mA	
v_o (mV)	--	677	
v_i (mV)	--	100	
i_g (nA)	--	90	
$A_v = v_o/v_s$ (V/V)	10	6.77	
$R_i = v_i/i_g$ (MΩ)	1.7	1.11	
R_o (kΩ)	2.6	2.467	

2. An NMOS transistor with an active load is shown in Fig. 6.4. For the 2N6784 NMOS, $V_{to} = 3$ V, $V_M = \infty$, and $K_P = 0.65$ A/V^2, and for the 2N6840 PMOS, $V_{to} = -3$ V, $V_M = \infty$, and $K_P = 1.85$ A/V^2.

Figure 6.4
An NMOS amplifier with an active load

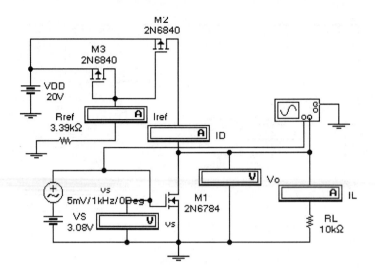

A. Open file FIG6_4.CA4 from the EWB file menu.
B. Check that the input signal v_s is 5 mV, 1 kHz. Otherwise, change the settings. Run the simulation.
C. Set the voltmeters for v_o and v_s to AC by double-clicking on the meters.
D. Check that the Oscilloscope has settings of AC; time base 0.1 ms/div; channel A: 5mV/div; channel B: 5 V/div. Otherwise, change the settings to give a clear display.
E. Zoom the oscilloscope display. Use the cursor to read v_o and v_s. Find the voltage gain.

        (7.083 V)   (7.073 mV)   (1001)
$V_{o(peak)}$ _____ $V_{s(peak)}$ _____ $A_v = V_{o(peak)}/V_{s(peak)}$ _____

F. Record $v_{o(ac)}$, $v_{s(ac)}$, the DC biasing current I_D and the DC reference current I_{ref} on Table 6.4.
G. Complete Table 6.4 by calculating the DC biasing current I_{CQ}, the DC reference current Iref, and the small-signal voltage gain Av. The small-signal voltage gain can be found from

$$A_V = -g_m(r_{o1} \| r_{o2} \| R_L) \tag{6-13}$$

where
$g_m = 2\,Kp\,(V_{GS} - V_{to})$ is the transconductance of transistor M_1.
$r_{o1} = V_{M1}/I_D$ is the output resistance of transistor M_1.
$r_{o2} = V_{M2}/I_D$ is the output resistance of transistor M_2.
$V_{M1} = 4$ for the NMOS transistor 2N67844, and $V_{M2} = 4$ for the PMOS transistor 2N6840.

Table 6.4

	Calculated	Simulated	Practically Measured
$V_{o(ac)}$ (V)		5.13V	
$V_{s(ac)}$ (mV)		5.00mV	
$A_v = V_{o(ac)}/V_{s(ac)}$		1026	
I_D (mA)	5	5.00mA	
I_{ref} (mA)	5	5.00mA	

6.8 DESIGN PROBLEMS

There may be more than one solution to the following problems. Use EWB or PSPICE to verify your design. Also, build and test, if possible. Determine the voltage and current ratings of active and passive components.

1. An amplifier with a transconductance gain of G_m = 20 mA/V±2% is required. The source resistance is R_s = 1 kΩ and the load resistance is R_L = 200 Ω. Determine the design specifications of the amplifier (input resistance, output resistance, and gain).

2. Design a JFET-amplifier, shown in Fig. 6.2, to give a mid-frequency voltage gain of $|A_v| = v_L/v_s = 20$ at a load resistance of R_L = 10 kΩ. Assume a source resistance of R_s = 500 Ω, and V_{DD} = 15 V.

3. Design a JFET-amplifier, shown in Fig. 6.3, to give a mid-frequency voltage gain of $|A_v| = |v_L/v_s| = 20$ at a load resistance of R_L = 10 kΩ. Assume a source resistance of R_s = 1.5 kΩ, and V_{DD} = 15 V.

4. Design a MOSFET-amplifier to give a mid-frequency voltage gain of $|A_v| = v_L/v_s = 10$ at a load resistance of R_L = 10 kΩ. Assume a source resistance of R_s = 1.5 kΩ, and V_{DD} = 12 V.

5. Design an NMOS amplifier, shown in Fig. 6.4, with an active load to give a mid-frequency voltage gain of $|A_v| = v_L/v_s \geq 250$ at a load resistance of R_L = 20 kΩ. Assume a source resistance of R_s = 1.5 kΩ, and V_{DD} = 15 V.

7 DESIGN OF BUFFER AMPLIFIERS

7.1 LEARNING OBJECTIVES

To design common-collector (BJT) and common-drain (FET) amplifiers to give a unity voltage gain, but a low output resistance and a high input resistance. We will use Electronics Workbench to verify the design and evaluate the performance of the amplifiers.

At the end of this lab, you will
- Be familiar with the operation and characteristics of common-collector and common-source amplifiers.
- Be able to analyze and design common-collector and common-source amplifiers to meet certain specifications.

7.2 THEORY

The common-collector and common-drain amplifiers give a unity gain, but offer a high input resistance and a low output resistance. They are commonly used as a buffer stage between a low impedance load and a high impedance source and are capable of supplying a low current. These amplifiers are often known as *buffer amplifiers*. Like the common-emitter (BJT) and common-source (FET) amplifiers, the buffer amplifiers must be biased properly to establish a quiescent point. The input signal is then superimposed on the quiescent gate (or base) voltage through coupling capacitors, or connected directly for direct-coupled amplifiers.

Common-Collector Amplifier

In a common-collector amplifier, the collector terminal is connected directly to the positive DC supply for an npn-transistor (or the negative supply for a pnp-transistor). A common-collector amplifier with a resistive biasing circuit is shown in Fig. 7.1. Resistors R_{B1} and R_{B2} set the base voltage V_B and hence the base current I_B, which, in turn, sets the collector current I_C. The emitter voltage V_E, which equals the output voltage, becomes

$$V_o = R_E I_E = V_B - V_{BE}$$

Since the base-emitter voltage V_{BE} is typically 0.7 V, $V_E = V_B - 0.7$ V and the voltage gain is almost unity, typically 0.96. There is no-phase shift between the input and the output voltages. Since V_E is V_B less V_{BE}, the output is shifted towards the zero (or the negative supply) and the circuit is also used as a *level shifter*.

The input signal v_s is superimposed on the base voltage, which causes the base current to change by a small amount. The change in the collector current is magnified due to the current gain of the transistor. However, the change in the emitter voltage is not magnified, because $v_E = v_B - v_{BE}$. Capacitor C_2 blocks any DC flowing to the load resistance, and the change of the emitter voltage will appear across the load. Thus, the output voltage v_o will be equal to the input signal v_s, and is in phase with it. Since the output (emitter) voltage follows the input voltage, this amplifier is often known as the *emitter follower*.

Figure 7.1
Emitter follower

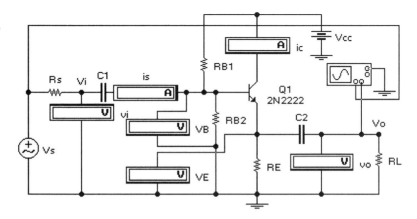

Resistors R_{B1} and R_{B2} can be replaced by a Thevenin's equivalent voltage V_{Th} and resistance R_{Th}. That is,

$$V_{Th} = \frac{R_{B2}}{R_{B1} + R_{B2}} V_{CC} \tag{7-1}$$

$$R_{Th} = \frac{R_{B1} R_{B2}}{R_{B1} + R_{B2}} \tag{7-2}$$

For a known value of V_{BE}, which is typically 0.7 V, the base current I_B can be found from,

$$I_B = \frac{V_{Th} - V_{BE}}{R_{Th} + (1 + \beta_F) R_E} \tag{7-3}$$

where β_F is the forward current gain of the transistor. The emitter current I_E can be found from

$$I_E = (1 + \beta_F) I_B \cong \beta_F I_B \tag{7-4}$$

Once the values of I_B and I_E are determined, then V_{CE} can be determined from

$$V_{CE} = V_{CC} - R_E I_E \tag{7-5}$$

The small-signal base-emitter resistance r_B of the transistor is given by

$$r_\pi = \frac{\beta_F V_T}{I_C} \tag{7-6}$$

where V_T = 25.8 mV. The small-signal transconductance g_m of the transistor is

$$g_m = \frac{\beta_F}{r_\pi} = \frac{I_C}{V_T} \tag{7-7}$$

The input resistance R_i of the amplifier is given by

$$R_i = \frac{v_i}{i_s} = R_B \| [r_\pi + (1+\beta_F)(R_E \| R_L)] \tag{7-8}$$

where $R_B = R_{B1} \| R_{B2}$.

The output resistance R_o of the amplifier is given by

$$R_o = R_E \left\| \frac{r_\pi + (R_s \| R_B)}{1+\beta_F} \right. \tag{7-9}$$

$$= (r_\pi + R_S)/\beta_F \quad \text{for } \beta_F \gg 1 \text{ and } R_S \ll R_B \tag{7-10}$$

The open-circuit voltage gain A_{vo} gain is given by

$$A_{vo} = \frac{v_o}{v_i} = \frac{(1+g_m r_\pi) R_E}{r_\pi + (1+\beta_F) R_E} = \frac{(1+\beta_F) R_E}{r_\pi + (1+\beta_F) R_E} \tag{7-11}$$

The *effective voltage gain* A_v is given by

$$A_v = \frac{v_o}{v_s} = \frac{A_{vo} R_i R_L}{(R_i + R_s)(R_L + R_o)} = \frac{A_{vo}}{(1 + R_s/R_i)(1 + R_o/R_L)} \tag{7-12}$$

Common-Drain Amplifier

A common-drain (JFET) amplifier with resistive biasing circuit is shown in Fig. 7.2. Resistors R_{G1} and R_{G2} set the gate voltage V_G, and hence the gate-source voltage V_{GS}, which, in turn, sets the drain current I_D. The source resistor R_{sr} adjusts the gate-source voltage V_{GS}. Resistor R_D sets mostly the drain-source voltage V_{DS}. The input signal v_s is superimposed on the gate voltage which causes the gate-source voltage to change by a small amount. The change in the drain current is magnified due to the transconductance gain of the transistor. As a result, the source voltage v_{SR} increases because $v_{SR} = v_G - v_{GS}$. The increase in v_{SR} equals the increase in v_G. Capacitor C_2 blocks any DC flowing to the load resistance and the change of the source voltage will appear across the load. Thus, the output voltage v_o is equal to the input signal v_s, and also in phase. Since the output (source) voltage follows the input voltage this amplifier is often known as the *source follower*.

Since the gate current of the JFET J_1, is very small, tending to be zero, the gate voltage follows the potential divider rule. That is,

$$V_G = \frac{R_{G2}}{R_{G1} + R_{G2}} V_{DD} \tag{7-13}$$

The gate-source voltage V_{GS} is given by

$$V_{GS} = V_G - R_{sr}I_D \qquad (7\text{-}14)$$

The gate-source voltage V_{GS} and the drain current I_D are related by

$$I_D = I_{DSS}\left(1 - \frac{V_{GS}}{V_P}\right) \qquad (7\text{-}15)$$

$$= K_p(V_{GS} - V_{to})^2 \qquad (7\text{-}16)$$

where I_{DSS} is the drain current for $V_{GS} = 0$ V, V_P is the pinch-off voltage of a JFET, $V_{to} = V_P$ for a JFET, and K_P is the FET constant; it is equal to I_{DSS}/V_P^2 for a JFET. Once the value of I_D is determined, then V_{DS} can be determined from

$$V_{DS} = V_{DD} - R_{sr}I_D \qquad (7\text{-}17)$$

Figure 7.2
Source follower

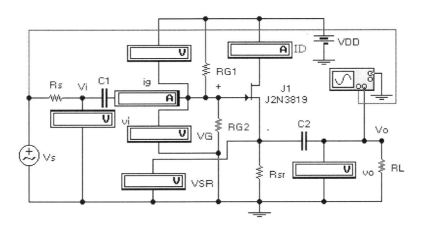

The small-signal transconductance g_m of the transistor is given by

$$g_m = 2K_p(V_{GS} - V_{to}) \qquad (7\text{-}18)$$

The input resistance R_i of the amplifier is given by

$$R_i = \frac{v_i}{i_g} = R_G = R_{G1}\|R_{G2} \qquad (7\text{-}19)$$

The output resistance R_o of the amplifier is given by

$$R_o = \frac{1}{g_m}\|R_{sr} = -\frac{R_{sr}}{1 + g_m R_{sr}} \qquad (7\text{-}20)$$

The open-circuit voltage gain A_{vo} gain is given by

$$A_{vo} = \frac{v_o}{v_i} = \frac{g_m R_{sr}}{1 + g_m R_{sr}} \qquad (7\text{-}21)$$

7.3 READING ASSIGNMENT

Study chapter(s) on emitter and source followers.

7.4 ASSIGNMENT

Design of Emitter-Follower Design the emitter follower shown in Fig. 7-1 to meet the following specifications:

- Voltage gain, $|A_v| = v_o/v_s \approx 1$.
- Load resistance $R_L = 10$ kΩ.
- Source resistance, $R_s = 200$ Ω.
- Input signal voltage, $v_s = 2$ V, 1 kHz.
- DC supply, $V_{CC} = 20$ V.
- Use NPN-transistor 2N2222 whose $\beta_F = 200$.
- Quiescent collector current, $I_C = 2$ mA (assumed).

BIASING RESISTORS

$r_\pi = \beta_F V_T/I_C = 200 \times 25.8$ mV/2 mA $= 2.58$ kΩ.
$\alpha_F = \beta_F/(1 + \beta_F) = 200/(1 + 200) = 0.995$.
$I_E = I_C/\alpha_F = I_C/0.995 = 2$ mA/0.995 $= 2.01$ mA.
Let us choose $V_E = V_{CC}/2 = 20/2 = 10$ V.
$V_B = V_{Th} = V_E + 0.7 = 10.7$ V.
$R_E = V_E/I_E = 10/2.01$ mA $= 5$ kΩ.
Make $(1 + \beta_F) R_E$ much larger than R_{th}. That is,
$R_{Th} = (1 + \beta_F) R_E/10 = (1 + 200) \times 5k/10 = 100.5$ kΩ
$R_{B1} = R_{Th} V_{CC}/V_{Th} = 100.5$ k $\times 20/10.7 = 188$ kΩ.
$R_{B2} = R_{Th} V_{CC}/(V_{CC} - V_{Th}) = 100.5$ k $\times 20/(20 - 10.7) = 216$ kΩ.
$R_i = R_B || [r_\pi + (1 + \beta_F)(R_E || R_L)] = 87.43$ Ω.

CAPACITORS

Capacitors C_1 and C_2 set the low-cut-off frequencies and usually have higher values. In Chapter 9, we will cover frequency response of amplifiers. Let us choose $C_1 = C_2 = 10$ µF so that they are effectively short-circuited at the frequency of the input signal.

Design of Source-Follower Design the source follower shown in Fig. 7.2 to meet the following specifications:

- Voltage gain, $|A_v| = v_o/v_s \equiv 1$.
- Load resistance, $R_L = 10$ kΩ.
- Source resistance, $R_s = 200$ Ω.
- Input signal voltage, $v_s = 2$ V, 1 kHz.

- DC supply, $V_{DD} = 20$ V.
- Use JFET transistor J2N3919 whose $I_{DSS} = 11$ mA, $V_p = -2.5$ V, and $K_p = 1.76$ mA/V².
- Quiescent drain current, $I_D = 2$ mA (assumed).

BIASING RESISTORS

The desired gate-source voltage can be found from
$I_D = K_p (V_{GS} - V_{to})^2$
which, for $K_{to} = V_p$, $K_p = 1.76$ mA/V² and $I_D = 2$ mA, gives $V_{GS} = -1.434$ V.

$g_m = 2 K_p (V_{GS} - V_{to}) = 2 \times 1.76$ mA/V $\times (-1.434 + 2.5) = 3.75232$ mA/V.
Let us choose $V_{SR} = V_{DD}/2 = 20/2 = 10$ V. Thus,

$$V_G = V_{SR} + V_{GS} = 10 - 1.434 = 8.566 \text{ V}$$

From Eq. (7-13), we can find R_{G1} and R_{G2}

$$\frac{R_{G1}}{R_{G2}} = \left(\frac{V_{DD}}{V_G} - 1\right) = 20/8.566 - 1 = 1.3348$$

Let $R_{G1} = 2.7$ MΩ, which gives $R_{G2} = 3.6$ MΩ. Thus,

$$V_G = 20 \times 2.7 \text{ M}\Omega/(2.7 \text{ M}\Omega + 3.6 \text{ M}\Omega) = 8.5714 \text{ V}$$
$$V_{SR} = V_G - V_{GS} = 8.5714 + 1.434 = 10.0054 \text{ V}$$

which gives

$$R_{SR} = V_{SR}/I_D = 10.0054 / 2 \text{ mA} = 5 \text{ k}\Omega$$

CAPACITORS

Let us choose $C_1 = C_2 = 10$ µF so that they are effectively short-circuits at the frequency of the input signal.

7.5 EWB SIMULATION

The EWB has circuit models for many types of BJTs and FETs. First, you get the schematic of a BJT from the list of Active devices and an FET from the list of FET devices. Then, you double-click on the device (BJT or FET) and the library of BJTs (or FETs) will open. The library has models of many commercially available BJTs and FETs. Also, you can change the model parameters by choosing **Edit**.

We will use EWB to simulate both the biasing circuit and the amplifier to verify the design specifications.

Emitter-Follower Circuit

The steps to follow are:
1. Open file FIG7_1.CA4 from the EWB file menu.
2. Check that the input signal is 2V, 1 kHz. Otherwise, change the settings.
3. Check that the meters for I_C, V_B, and V_E are set to DC, and the meters for v_o, v_i, and i_g are set to AC. Double-click the meter and then change, if needed.
4. Run the simulation. Record the values in Table 7.1.
5. To find the output resistance R_o, run the simulation for R_L = 10 kΩ, 10 MΩ.

$$R_o = R_L(=10\ k\Omega)\left[\frac{v_o(\text{for } R_L = 10\ M\Omega)}{v_o(\text{for } R_L = 10\ k\Omega)} - 1\right] \quad (7\text{-}22)$$

R_o, = 10 kΩ. x (1.99V/1.98V - 1) = 50.5 Ω.

6. Check that the Oscilloscope has setting of AC (if it comes up DC, then change to AC); time base: 0.1 ms/div; channel A: 2 V/div; channel B: 1 V/div. Otherwise, change the settings to give a clear display.
7. Zoom the oscilloscope display. Use the cursor to read the vo and vs. Find the voltage gain.

    (2.8043 V)   (2.8281 V)   (0.9916)

$V_{o(peak)}$ _____ $V_{s(peak)}$ _____ $A_v = V_{o(peak)}/V_{s(peak)}$ _____

8. Compare the input and output waveforms. Comment.

_____

_____

_____

9. If you change the load resistance R_L, would the input resistance R_i change? Why?

_____

_____

_____

10. If you change the source resistance R_s, would the output resistance R_i change? Why?

_____

_____

_____

Table 7.1

	Calculated	Simulated	Practically Measured
I_C (mA)	2	1.68	
V_B (V)	10.7	9.02	
V_E (V)	10	8.38	
v_i (V)	--	1.99	
v_o	--	1.98V	
i_s (μA)	--	24.7μA	
$A_V = v_o/v_i$	1	0.995	
$R_i = v_i/i_s$ (kΩ)	87.43	80.57 kΩ	
R_o (Ω)	13.8	50.5Ω	

Source-Follower Circuit

The steps to follow are:
1. Open file FIG7_2.CA4 from the EWB file menu.
2. Check that the input signal is 2V, 1 kHz. Otherwise, change the settings.
3. Check that the meters for I_D, V_G and V_{SR}, are set to DC and the meters for v_o, v_i and i_g are set to AC. Double-click the meter and then change, if needed.
4. Run the simulation. Record the values in Table 7.2.
5. To find the output resistance Ro. Run the simulation for RL = 10 kΩ, 10 MΩ.

$$R_o = R_L(=10 \text{ k}\Omega)\left[\frac{v_o(\text{for } R_L = 10 \text{ M}\Omega)}{v_o(\text{for } R_L = 10 \text{ k}\Omega)} - 1\right] \quad (7\text{-}23)$$

R_o, = 10 kΩ. x (1.89V/1.85 V - 1) = 216.2 Ω.

6. Check that the Oscilloscope has settings of AC (if it is in DC, change to AC); time base 0.2 ms/div; channel A: 1 V/div; channel B: 1 V/div. Otherwise, change the settings to give a clear display.
7. Zoom the oscilloscope display. Use the cursor to read the v_o and v_s. Find the voltage gain.

(2.6224 V)   (2.8281 V)   (0.9273)
$V_{o(peak)}$ _____ $V_{s(peak)}$ _____ $A_v = V_{o(peak)}/V_{s(peak)}$ _____

8. Compare the input and output waveforms. Comment.

_____
_____
_____

9. If you change the load resistance R_L, would the input resistance R_i change? Why?

_____
_____
_____

10. If you change the source resistance R_s, would the output resistance R_i change? Why?

_____
_____
_____

11. Compare the output voltage waveform of the emitter follower with that of the source follower. Comment on the difference if any.

_____
_____

Table 7.2

	Calculated	Simulated	Practically Measured
I_D (mA)	2	2.23	
V_G (V)	8.5714	9.65	
V_{SR} (V)	10.00554	11.1	
v_i (V)	--	2.00	
v_o (V)	--	1.85	
i_g (μA)	--	5.29	
$A_V = v_o/v_s$	1	0.925	
$R_i = R_s v_i/(v_s - v_i)$ (MΩ)	1.54	∞, since $v_s = v_i$	
R_o (Ω)	253	216	

7.6 WRITE-UP/CONCLUSIONS

Write a brief report, summarizing the results of the design experiment and what you have learned or confirmed about the buffer amplifiers.

Comment on the relationship between theory, simulation and practical circuit performance and the precautions that must be taken when using this learning approach.

_____
_____
_____
_____
_____
_____
_____

7.7 REINFORCEMENT EXERCISES

An emitter follower with an active load is shown in Fig. 7.3. For the npn-transistor 2N3904, Early voltage $V_A = 113$ V, and $\beta_F = 204$.
1. Open file FIG7_3.CA4 from the EWB file menu.
2. Check that the input signal is 2V, 1 kHz. Otherwise, change the settings.
3. Check that the meters for I_C, and I_{ref} are set to DC, and the meter for v_o, to AC. Double-click the meter and then change, if needed.
4. Run the simulation. Record the values in Table 7.3.

Figure 7.3
Emitter follower with an active load

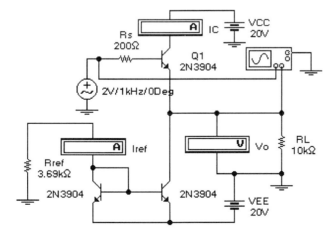

Table 7.3

	Calculated	Simulated	Practically Measured
I_C (mA)	5.23	5.82	
I_{ref} (V)	5.23	-5.22	
v_o (V)	2	2.00	
v_s (V)	2	2.00	
$A_V = v_o/v_s$	1	1.0	

5. Check that the Oscilloscope has setting of AC; time base: 0.2 ms/div; channel A: 1 V/div; channel B: 1 V/div. Otherwise, change the settings to give a clear display.

6. Zoom the oscilloscope display. Use the cursor to read v_o and v_s. Find the voltage gain.

7.   (2.0777 V)   (2.8284 V)   (0.735)

$V_{o(peak)}$ _____ $V_{s(peak)}$ _____ $A_v = V_{o(peak)}/V_{s(peak)}$ _____

7. Check that the Oscilloscope has settings of DC; time base: 0.1 ms/div; channel A: 1 V/div; channel B: 1 V/div. Otherwise, change the settings to give a clear display. The output voltage is shifted down? Why? Explain.

_____

_____

_____

8. Compare the output voltage waveforms of this circuit with that of the emitter follower in Fig.7.1. Comment on the differences.

_____

_____

_____

9. Complete Table 7.3 by calculating the DC biasing current I_C, the DC reference current I_{ref}, and the small-signal voltage gain A_V. The small-signal voltage gain can be found from

$$A_v = \frac{v_o}{v_i} = \frac{(1+\beta_F)(r_{01}\|r_{02}\|R_L)}{r_\pi + (1+\beta_f)(r_{01}\|r_{02}\|R_L)} \quad (7\text{-}24)$$

$r_{01} = V_{A1}/I_C$ is the output resistance of transistor Q_1.
$r_{02} = V_{A2}/I_C$ is the output resistance of transistor Q_2.
$V_{A1} = V_{A2} = V_A = 204$ V for the npn-transistor 2N3904.
$\beta_F = 100$ for the npn-transistor 2N3904.

7.8 DESIGN PROBLEMS

There may be more than one solution to the following problems. Use EWB or PSPICE to verify your design. Also, build and test, if possible. Determine the voltage and current ratings of active and passive components.

1. Design a BJT-buffer-amplifier, shown in Fig. 7.1, to give a mid-frequency voltage gain of $|A_v| = |v_L/v_s| \approx 1$, with an input resistance of $R_{in} = v_s/i_s \geq 50$ kΩ. The load resistance is $R_L = 10$ kΩ. Assume a source resistance of $R_s = 500$ Ω and $V_{CC} = 15$ V.

2. Design an FET-buffer-amplifier, shown in Fig. 7.2, to give a mid-frequency voltage gain of $|A_v| = |v_L/v_s| \approx 1$, and an input resistance of $R_{in} = v_s/i_s \geq 500$ kΩ. The load resistance is $R_L = 10$ kΩ. Assume a source resistance of $R_s = 500$ Ω and $V_{DD} = 15$ V.

3. Design a BJT-buffer-amplifier, shown in Fig. 7.3, with an active load to give a mid-frequency voltage gain of $|A_v| = |v_L/v_s| \approx 1$, and an input resistance of $R_{in} = v_s/i_s \geq 50$ kΩ. The load resistance is $R_L = 10$ kΩ. Assume a source resistance of $R_s = 500$ Ω, and $V_{CC} = -V_{EE} = 15$ V.

4. The input signal $v_s = 2$ mV to an amplifier with a source resistance of $R_s = 1$ kΩ. Design a BJT-amplifier to give a mid-frequency voltage gain $A_v (= v_L/v_s$ with a load resistance of $R_L = 10$ kΩ) of greater than 650. The input resistance R_i of the amplifier should be greater than 70 kΩ and the output resistance R_{out} should be less than 250 Ω. The DC supply voltage is $V_{CC} = 12$ V.

5. The input signal $v_s = 2$ mV to an amplifier with a source resistance of $R_s = 1$ kΩ. Design an FET amplifier to give a mid-frequency voltage gain $A_v (= v_L/v_s$ with a load resistance of $R_L = 10$ kΩ) of greater than 450. The input resistance R_i of the amplifier should be greater than 500 kΩ, and the output resistance R_{out} should be less than 250. The DC supply voltage is $V_{DD} = 12$ V.

8 DESIGN OF POWER AMPLIFIERS

8.1 LEARNING OBJECTIVES

To design a class-AB power amplifier to give an output voltage without any distortion at unity voltage gain. We will use Electronics Workbench to verify the design and evaluate the performance of the amplifier.

At the end of this lab, you will
- Be familiar with the operation and characteristics of class-B and class-AB amplifiers.
- Be able to analyze and design a class-AB amplifier to meet certain specifications.

8.2 THEORY

The amplifiers, which we discussed in Chapters 5, 6, and 7, are of class-A type. In a class-A amplifier, the DC biasing collector current I_C of a transistor is higher than the peak amplitude of the AC output current, I_p. Thus, the transistor in a class-A amplifier conducts during the entire cycle of the input signal, and the conduction angle is $\Theta = \omega t = 360°$. The efficiency of this type of amplifier is low, typically 25%, and it introduces distortion of the output waveform.

In a class-B amplifier, the transistor is biased at zero DC current, and it conducts for only one half-cycle of the input signal with a conduction angle of $\Theta = 180°$. The negative half of the sinusoid is provided by another transistor that also operates in the class-B mode and conducts during the alternate half-cycles. Two complementary transistors (i.e., npn and pnp transistors) are employed to perform the push-pull class-B operation. A class-B amplifier is shown in Fig. 8.1. If the input voltage is positive, the npn-transistor Q_N conducts and behaves like an emitter follower. If the input voltage is negative, the pnp-transistor Q_P conducts and behaves like an emitter follower. Thus, each transistor carries the peak load current, and the efficiency can be as high as 78%.

Figure 8.1
Class-B amplifier

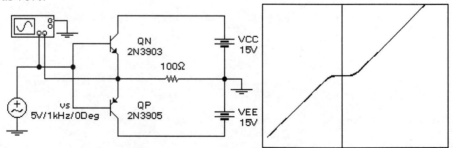

If the input voltage is less than the minimum voltage to turn on a transistor, there will be no output, and the output voltage will be distorted. The transfer characteristic (v_o versus v_s) shows a dead zone near the zero-crossing of the input voltage, and this will introduce cross-over distortion in the output voltage. This arrangement is also known as the class-B *push-pull* amplifier.

In order to eliminate this distortion, the amplifier is normally operated in class-AB operation. In a class-AB amplifier, the transistor is biased at a nonzero DC current that is much smaller than the peak amplitude of the load current. A DC biasing voltage V_{BB} is applied between the bases of Q_N and Q_P so that a small quiescent biasing current of I_C flows continuously through the transistors. For $v_s = 0$, a voltage $V_{BB}/2$ appears across the base-emitter junction of both Q_N and Q_P. By choosing $V_{BB}/2 = V_{BEN} = V_{EBP}$, both transistors will be on the verge of conduction. That is, $v_o = 0$ for $v_s = 0$. Only a small positive input voltage v_s will then cause Q_N to conduct, and similarly a small-negative voltage will cause Q_P to conduct. The arrangement of a class-AB amplifier is shown in Fig. 8.2. The two diodes D_1 and D_2 (together with R1 and R_2) provide the biasing voltage V_{BB}.

Figure 8.2
Class-AB amplifier

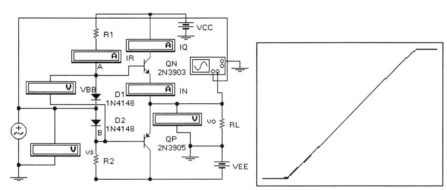

As v_s becomes positive, Q_N acts as an emitter follower delivering output power, with Q_P conducting only a very small current. When v_s becomes negative, the opposite occurs and Q_P acts as an emitter follower. Thus, v_o follows the input signal v_s. The circuit operates in class-AB mode, because both transistors remain on and operate in the active region. The currents from the two transistors are combined to form the load current. Both transistors conduct for an interval near the zero crossings of the input signal. The transfer characteristic (v_o versus v_s) shows the elimination of crossover distortion. This arrangement is also known as the class-AB push-pull amplifier. The output voltage v_o is given by

$$v_o = v_s + V_{BB}/2 - V_{BEN}(=V_{EBP}) \tag{8-1}$$

which, for identical transistors of $V_{BEN} = V_{EBP}$, and $V_{BB}/2 = V_{BEN}$, gives $v_o = v_s$. For this reason, most of the crossover distortion is eliminated.

For a positive v_o, a current i_o flows through R_L. That is, the current through Q_N is

$$i_N = i_P + i_o \quad (8\text{-}2)$$

Any increase in i_N will cause a corresponding increase in V_{BEN} above the quiescent value of $V_{BB}/2$. Since V_{BB} must remain constant, the increase in V_{BEN} will cause an equal decrease in V_{EBP} and hence in i_P. Thus,

$$V_{BB} = V_{BEN} + V_{EBP} \quad (8\text{-}3)$$

If expressed in terms of saturation current I_S, Eq. (8-3) becomes

$$2V_T \ln\left(\frac{I_Q}{I_S}\right) = V_T \ln\left(\frac{i_N}{I_S}\right) + V_T \ln\left(\frac{i_P}{I_S}\right)$$

which, after simplification, gives

$$I_Q^2 = i_N i_P \quad (8\text{-}4)$$

$$= i_N(i_N - i_o) = i_N^2 - i_N i_o \quad (8\text{-}5)$$

This can be solved for the current i_N for a given quiescent current I_Q. Thus, as i_N increases, i_P decreases by the same ratio. However, their product remains constant. Resistances R_1 and R_2 provide the quiescent current I_Q for the transistors and also ensure conduction of the diodes. That is,

$$I_R = I_{D1} + i_N/(1+\beta_F) \approx I_{D1} + (I_Q + i_o)/(1+\beta_F)$$

which must guarantee the base-biasing current for Q_N when the load current becomes maximum. Thus, the values of R_1 and R_2 can be found from

$$R_1 = R_2 = \frac{V_{CC} - V_{D1}(=V_{D2} = V_{BB}/2)}{I_{D1(min)} + (I_Q + i_{o(max)})/(1+\beta_F)} \quad (8\text{-}6)$$

where $I_Q = I_S \cdot \exp(V_{BB}/2V_T)$, and it is usually smaller than $i_{o(max)}$. Hence I_Q can often be neglected for finding the values of R_1 and R_2. $I_{D1(min)}$ is the minimum current to ensure diode conduction.

8.3 READING ASSIGNMENT

Study chapter(s) on power amplifiers.

8.4 DESIGN SPECIFICATIONS

Design the class-AB amplifier shown in Fig. 8.2 to meet the following specifications:

- Since it is a power amplifier, not a voltage amplifier, voltage gain, $|A_v| = v_o/v_s \equiv 1$.
- Load resistance, $R_L = 100\ \Omega$.
- Peak load current, $i_{o(peak)} = 150$ mA.
- Quiescent collector biasing current, $I_C = 2$ mA.
- Source resistance, $R_s = 0\ \Omega$.
- Input signal voltage, $v_s = 5$ V, 1 kHz.
- DC supply voltages, $V_{CC} = -V_{EE} = 15$ V.

Use NPN-transistor 2N3903 whose $\beta_{FN} = 102$, $V_A = 113$, and $I_S = 7.62\text{e--}16$ A, PNP-Transistor 2N3904 whose $\beta_{FP} = 100$, $V_A = 113$, and $I_S = 2.01\text{e--}15$ A, and Diodes 2N4148 whose $I_S = 1.42\text{e--}18$ A.

Biasing Resistors

The peak emitter current of the transistor Q_N is $i_{N(max)} \equiv i_P + i_{o(max)} = 2$ mA + 150 mA = 152 mA. The peak base current of the transistor Q_N is $i_{BN(max)} \equiv (I_Q + i_{o(max)})/\beta_N = 152$ mA $/102 = 1.49$ mA.

Let us assume that the diodes require a minimum of 1 mA to conduct. That is, $I_{D1(min)} = 1$ mA. Thus, the required value of biasing current through R_1 is

$$I_R = I_{D1(min)} + (I_Q + i_{o(max)})/(1 + \beta_{FN}) = 1\ \text{mA} + 1.49\ \text{mA} = 2.49\ \text{mA}.$$

The biasing voltage V_{BB} required to maintain a biasing current of I_Q, can be found from

$$V_{BB} = V_{BEN} + V_{EBP}$$

which is

$$V_T \ln\left(\frac{I_Q}{I_{SN}}\right) + V_T \ln\left(\frac{I_Q}{I_{SP}}\right) = 25.8\ \text{mV} \times \left[\ln\left(\frac{2\ \text{mA}}{7.62\text{e-}16}\right) + \ln\left(\frac{2\ \text{mA}}{2.01\text{e-}15}\right)\right] = 1.4506\ \text{V}$$

Thus, the values of R_1 and R_2 become

$$R_1 = R_2 = \frac{V_{CC} - V_{BB}/2}{I_R} = \frac{15 - 1.4506/2}{2.49\ \text{mA}} = 5.4\ \text{k}\Omega$$

8.5   EWB SIMULATION

The EWB has circuit models for many types of diodes and BJTs. First, you get the schematic of a diode and BJT from the list of Active devices. Then, you double-click on the device (BJT or diode) and the library of BJTs (or diodes) will open. The library has models of many commercially available diodes and BJTs. Also, you can change the model parameters by choosing **Edit**.

We will use EWB to simulate both the class-A amplifier in Fig. 8.1, and the class-AB amplifier in Fig. 8.2 to verify the design specifications.

Class-B Amplifier

The steps to follow are:
1. Open file FIG8_1.CA4 from the EWB file menu. Run the simulation.
2. Check that the input signal is 4V, 1 kHz. Otherwise, change the settings.
3. Check that the meters for v_o and v_s are set to AC. Double-click the meter and then change, if needed.
4. Check that the Oscilloscope has settings of AC (for both channels); time base: 0.1 ms/div and Y/T (signal versus time); channel A: 2 V/div; channel B: 2 V/div. Otherwise, change the settings to give a clear display.
5. Zoom the oscilloscope display. Use the cursor to read the v_o and v_s. Find the voltage gain.

    (4.8318 V)   (5.6586 V)   (0.8539)

$V_{o(peak)}$ _____ $V_{s(peak)}$ _____ $A_v = V_{o(peak)}/V_{s(peak)}$ _____

6. Record v_o and v_s from the voltmeters, and find the voltage gain.

    (3.29 V)   (4.0 V)   (0.8225)

V_o _____ V_s _____ $A_v = v_o/v_s$ _____

Does this gain differ from that obtained from the oscilloscope? If yes, why?

_____

_____

_____

7. Compare the input and output waveforms. You can see cross-over distortion. What is the minimum value of v_s which is required to produce an output voltage? $v_{s(min)}$ _____ (≈ 0.7 V)
8. Increase the value of the input voltage and carefully observe the resulting output voltage waveform. What are the findings?

_____

_____

_____

What are the maximum and minimum values of the sinusoidal output voltages without any clipping?

$V_{o(max)}$ _____ $V_{s(max)}$ _____ $V_{o(min)}$ _____ $V_{s(min)}$ _____

Class-AB Amplifier

The steps to follow are:
1. Open file FIG8_2.CA4 from the EWB file menu. Run the simulation.
2. Check that the input signal is 4V, 1 kHz. Otherwise, change the settings to give a good display.

3. Check that the meters for $I_R, V_{BB}, I_N,$ and I_Q are set to DC, and the meters for v_o, and v_s are set to AC. Double-click the meter and then change, if needed.

4. Check that the Oscilloscope has settings of DC; time base: 0.1 ms/div and Y/T (signal versus time); channel A: 2 V/div; channel B: 2 V/div. Otherwise, change the settings to give a good readable display.

5. Zoom the oscilloscope display. Use the cursor to read the v_o and v_s. Find the voltage gain.
$v_{o(peak)}$ _____ (5.5425 V) $v_{s(peak)}$ _____ (5.6589 V) $A_v = v_{o(peak)}/v_{s(peak)}$ = _____ (0.9798)

6. Record v_o and v_s from the voltmeters and find the voltage gain.
v_o _____ (3.93 V) v_s _____ (4.0 V) $A_v = v_o/v_s$ = _____ (0.9825)
Is this gain close to that obtained from the oscilloscope? If yes, why?

_____

_____

_____

7. Compare the input and output waveforms. Do you see crossover distortion? _____ (No). What is the minimum value of v_s that is required to produce an output voltage? $v_{s(min)}$ _____ ($\approx$ oV).

8. Increase the value of the input voltage and carefully observe the resulting output voltage waveform. What are the maximum and minimum values of the sinusoidal output voltages without any clipping?
$v_{o(max)}$ _____ $v_{s(max)}$ _____ ($\approx$ 5.5 V) $v_{o(min)}$ _____ $v_{s(main)}$ _____ (0 V)

9. Change the input signal to 15V at 1 kHz. Also, change the time base to B/A (channel B versus channel A). Run the simulation. You will see the transfer function (v_o versus v_s). Record.
$v_{o(max)}$ _____ (7.228 V) $v_{s(max)}$ _____ (7.232 V) $v_{o(min)}$ _____ (-7.5014 V)
$v_{s(min)}$ _____ (-7.5064 V)
Observe the transfer characteristic carefully. What are the findings?

_____

_____

_____

10. (Only for EWB version 5.0) Compare the output voltage waveform of the class-AB amplifier with that of the class-B amplifier. Comment on the differences, if any.

_____

_____

_____

11. Run the Fourier analysis of the output voltage for the class-B amplifier (at node 3) in Fig. 8.1 and the class-AB amplifier (at node 4) in Fig. 8.2. Set output variable at node = 3 for class A and 4 for class AB, fundamental frequency = 1 kHz, and number of harmonics = 9. Compare the total harmonic distortions (THD): THD_A _____ and THD_{AB} _____.
Comment on the results:

_____

_____

_____

8.6 WRITE-UP/CONCLUSIONS

Write a brief report, summarizing the results of the design experiment and what you have learned or confirmed about the class-A and class-AB amplifiers.

Comment on the relationship between theory, simulation and practical circuit performance and the precautions that must be taken when using this learning approach.

_____

_____

_____

_____

_____

_____

_____

8.7 REINFORCEMENT EXERCISES

A class-AB amplifier with an active load is shown in Fig. 8.3.
1. Open file FIG8_3.CA4 from the EWB file menu. Run the simulation.

Figure 8.3
Class-AB amplifier with an active load

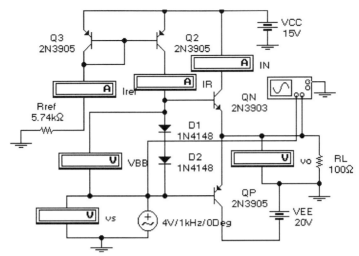

2. Check that the input signal is 4V, 1 kHz. Otherwise, change the settings.
3. Check that the meters for I_{ref}, I_R, I_N and V_{BB} are set to DC, and the meters for v_o, and v_s to AC. Double-click the meter and then change, if needed.
4. Check that the Oscilloscope has settings of DC; time base: 0.1 ms/div and Y/T (signal versus time);; channel A: 2 V/div; channel B: 2 V/div. Otherwise, change the settings to give a clear display.
5. Zoom the oscilloscope display. Use the cursor to read the v_o and v_s. Find the voltage gain.

$V_{o(peak)}$ _____ (6.452 V) $v_{s(peak)}$ _____ (5.657 V) $A_v = v_{o(peak)}/v_{s(peak)} =$

_____ (1.1405)

6. Record v_o and v_s from the voltmeters and find the voltage gain.

v_o _____ (3.96 V) v_s _____ (4.0 V) $A_v = v_o/v_s =$ _____ (0.99)

Is this gain close to that obtained from the oscilloscope? If yes, why?

_____
_____
_____

7. Record the voltages and currents and complete Table 8.1.

Table 8.1

	Calculated	Simulated	Practically Measured
I_{Ref} (mA)	2.49	2.49	
I_R (mA)	2.49	2.69	
I_N (mA)	152	80.1	
V_{BB} (V)	1.45	1.72	
v_o (V)	4	3.96	
v_s (V)	4	4.00	
$A_V = v_o/v_s$	1	0.99	

8. Compare the input and output waveforms. Do you see crossover distortion? _____ (No). What is the minimum value of v_s that is required to produce an output voltage? $v_{s(min)}$ _____ . The output voltage is shifted upward. Why? Explain.

_____

_____

_____

9. Increase the value of the input voltage and carefully observe the resulting output voltage waveform. What are the maximum and minimum values of the sinusoidal output voltages without any clipping?
$v_{o(max)}$ _____ (≈ 7 V) $v_{s(max)}$ _____ $v_{o(min)}$ _____ $v_{s(main)}$ _____

10. Change the input signal to 15V at 1 kHz. Also, change the time base to B/A. Run the simulation. You will see the transfer function (v_o versus v_s). Record.
$v_{o(max)}$ _____ (1.0833 V) $v_{s(max)}$ _____ (1.0838 V) $v_{o(min)}$ _____ (-1.9764 V) $v_{s(min)}$ _____ (-1.9764 V)
Observe the transfer characteristic carefully. What are the findings?

_____

_____

_____

8.8 DESIGN PROBLEMS

There may be more than one solution to the following problems. Use EWB or PSPICE to verify your design. Also, build and test, if possible. Determine the voltage and current ratings of active and passive components.

1. Design a BJT class-AB amplifier, shown in Fig. 8.2, to supply a maximum output power of $P_{L(max)} = 20$ W at a load resistance of $R_L = 50\,\Omega$. Assume a DC supply voltage of $V_{CC} = -V_{EE} = 15$V.

2. Design a BJT class-AB amplifier, shown in Fig. 8.3, with an active biasing to supply a maximum output power of $P_{L(max)} = 20$ W at a load resistance of $R_L = 50\,\Omega$. Assume a DC supply voltage of $V_{CC} = -V_{EE} = 15$V.

9 DESIGN OF DIFFERENTIAL AMPLIFIERS

9.1 LEARNING OBJECTIVES

To design a differential amplifier to give a specified differential voltage gain. We will use Electronics Workbench to verify the design and evaluate the performance of the amplifier.

At the end of this lab, you will
- Be familiar with the operation and characteristics of differential amplifiers.
- Be able to analyze and design a differential amplifier to meet certain specifications.

9.2 THEORY

A differential amplifier consists of an emitter-coupled (or source-coupled) pair. The general configuration of a BJT differential amplifier is shown in Fig. 9.1. The two sides of the pair should be identical. That is, $R_{C1} = R_{C2}$, $I_{C1} = I_{C2}$, and $I_{B1} = I_{B2}$. Also, the transistors should be identical. The difference output voltage $v_{o(diff)} = 0$. In practice, mismatches do exist and there is a very small offset voltage. The biasing currents should be such that the transistors are operated in the active region. The DC biasing circuit, which is shown as a constant current source (with R_{EE} as its output resistance) can be either a simple resistor, in which case the equivalent current generator will be zero, or a transistor current source which is generally used in ICs. For measuring the DC biasing voltages and currents, both inputs are shorted to ground.

Figure 9.1
BJT differential amplifier

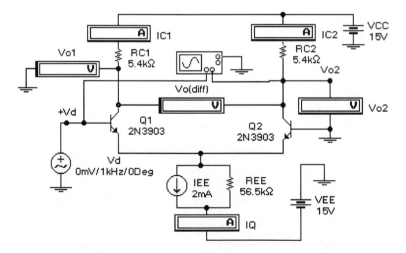

The performance of a differential amplifier is measured by a differential gain A_d in response to a difference voltage v_{id} between the two input terminals, a common mode gain A_c in response to a voltage v_{ic} common to both input terminals, and a common mode rejection ratio (CMRR). The CMRR is the ratio of the differential gain to the common mode gain, and it is a measure of the ability of an amplifier to amplify the differential signal while rejecting the common mode signal. Thus, a differential amplifier requires the design and analysis for (a) DC biasing, (b) small-signal differential input v_{id}, and (c) small-signal common mode input v_{ic}. The output voltage is found from

$$v_o = A_{id} v_{id} + A_c v_{ic} \tag{9-1}$$

DC Biasing of Differential Amplifier

Differential amplifier is generally biased with an active current source in order to produce a stable biasing current and also to give a high output resistance R_{EE}. A high value of R_{EE} is required to give a high CMRR. There are many types of current sources. The differential amplifier with a two-transistor current source is shown in Fig. 9.2. The biasing current I_Q, which is set by R_{ref} and V_{EE}, can be found from

$$I_Q = I_{ref} = \frac{V_{EE} - V_{BE}}{R_{ref}} \tag{9-2}$$

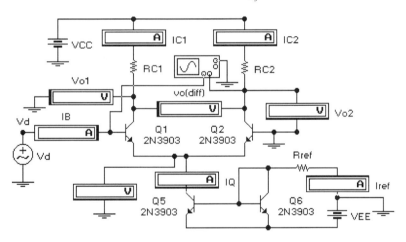

Figure 9.2
BJT differential amplifier with an active current biasing

The DC collector currents of the transistors are

$$I_{C1} = I_{C2} = I_Q/2 \tag{9-3}$$

The output resistance of the current source is given by

$$R_{EE} = \frac{V_{A5}}{I_Q} \tag{9-4}$$

where V_{A5} is the Early voltage of transistor Q_5.

Differential Mode Signal

The amplifier is operated with a small differential voltage near zero in the linear portion of the transfer characteristic. Let us assume that there is no common mode signal, that is, $v_{ic} = 0$, and only the differential input voltage v_d is applied. v_{id} can be applied to only one terminal with the other input terminal shorted to ground as shown in Fig. 9.3. Alternately, $v_{id}/2$ can be applied to one terminal and $-v_{id}/2$ to the other terminal so that the differential voltage is v_{id}. If v_{id} is applied to one terminal as shown, the base-emitter voltage of Q_1 will increase by $+v_{id}/2$ and that of Q_2 will decease by $-v_{id}/2$. Assuming that the two transistors are identical and the circuit is balanced, the increase in voltage at the emitter junction due to $+v_{id}/2$ will be compensated by an equal amount of decrease in voltage due to $-v_{id}/2$. As a result, the voltage at the emitters of the transistors will not vary at all. The emitter junction, which experiences no voltage variation, can be regarded as the ground potential. The capacitor C_2 is connected to block the flow of DC current to the load.

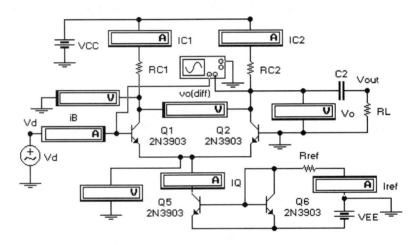

Figure 9.3
BJT differential amplifier with a differential input voltage

The small-signal differential mode voltage gain A_d with the voltage difference output can be found from

$$A_d = \frac{v_{o(diff)}}{v_{id}} = -g_m \left(R_C \| r_{o2} \right) \quad (9\text{-}5)$$

where r_{o2} is the output resistance of transistor Q_2.

The small-signal differential mode voltage gain A_d with the single-ended output can be found from

$$A_d = \frac{v_{od}}{v_{id}} = -\frac{g_m \left(R_C \| r_{o2} \right)}{2} \quad (9\text{-}6)$$

where $g_m = I_C/V_T$ is the transconductance of transistor Q_1 or Q_2.

The small-signal differential mode input resistance R_{id} can be found from

$$R_{id} = \frac{v_{id}}{i_d} = 2r_\pi \qquad (9\text{-}7)$$

where $r_\pi = \beta_F V_T/I_C$ is the base-emitter resistance of transistor Q_1 or Q_2.

Common Mode Signal
The differential amplifier with only common mode input v_{ic} is shown in Fig. 9.4. The same voltage v_{ic} is applied to both the terminals. Since the input voltages to both sides are the same, the collector voltages will change by the same amount. The emitter resistance R_{EE}, which has a large value, will influence the voltage gain. However, the gain will be low.

Figure 9.4
BJT differential amplifier with a common mode input voltage

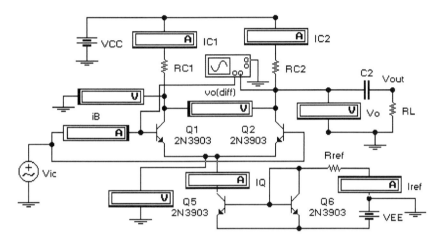

The small-signal common mode voltage gain A_c with the difference voltage output can be found from

$$A_c = \frac{v_{oc}}{v_{ic}} \equiv \frac{-g_m(R_C \| r_{02})}{1 + 2g_m R_{EE}(1 + 1/\beta_F)} \qquad (9\text{-}8)$$

The small-signal common mode voltage gain A_c with the single-ended output can be found from

$$A_c = \frac{v_{oc}}{v_{ic}} \equiv \frac{-(g_m/2)(R_C \| r_{02})}{1 + 2g_m R_{EE}(1 + 1/\beta_F)} \qquad (9\text{-}9)$$

The small-signal common mode input resistance R_{ic} can be found from

$$R_{ic} = \frac{v_{ic}}{i_b} = r_\pi + (1 + \beta_F) 2 R_{EE} \qquad (9\text{-}10)$$

Common Signal CMRR

The small-signal common-mode rejection ratio (CMRR) can be found from

$$\text{CMRR} = \left|\frac{A_d}{A_c}\right| = 1 + 2g_m R_{EE}\left(1 + \frac{1}{\beta_F}\right) \qquad (9\text{-}11)$$

$$= 2g_m R_{EE} \quad \text{for } \beta_F \gg 1, \text{ and } 2g_m R_{EE} \gg 1 \qquad (9\text{-}12)$$

9.3 READING ASSIGNMENT

Study chapter(s) on differential amplifiers.

9.4 DESIGN SPECIFICATIONS

Design the differential amplifier shown in Fig. 9.3 to meet the following specifications:
- No-load differential mode voltage gain, $|A_d| = v_{od}/v_{id} = 100$.
- Load resistance $R_L = 100$ kΩ.
- Quiescent biasing current, $I_Q = 2$ mA.
- Source resistance, $R_s = 0$ Ω.
- Differential input signal voltage, $v_{id} = 10$ mV, 1 kHz.
- Common-mode input signal voltage, $v_{ic} = 2$V, 1 kHz.
- DC supply voltages, $V_{CC} = -V_{EE} = 15$ V.

Use NPN-transistor 2N3903 whose $\beta_{FN} = 102$, $V_A = 113$, and $I_S = 7.62\text{e-}16$ A, and a PNP-transistor 2N3904 whose $\beta_{FN} = 100$, $V_A = 113$, and $I_S = 2.01\text{e-}15$ A.

Biasing Current Source

The value of R_{ref} can be found from

$$R_{ref} = \frac{V_{EE} - V_{BE}}{I_Q} = \frac{15 - 0.7}{2 \text{ mA}} = 7.2 \text{ k}\Omega$$

$$R_{EE} = r_{o5} = V_A/I_Q = 113/2 \text{ mA} = 56.5 \text{ k}\Omega$$

$r_{o2} = V_A/I_C = 113/1$ mA $= 113$ kΩ.
$g_m = I_C/V_T = 1$ mA$/25.8$ mV $= 38.76$ mA/V.

Collector Resistors

For a maximum output voltage swing, the DC voltage drop across the collector resistors R_{C1} and R_{C2} should be limited to $V_{CC}/2$. That is, $R_C I_C = V_{CC}/2$, which gives the maximum value of R_C as

$$R_{C(max)} = V_{CC}/2I_C = 15/(2 \times 1 \text{ mA}) = 7.5 \text{ k}\Omega$$

For the single-ended output, A_d is given by

$$A_d = -\frac{g_m(R_C \| r_{02})}{2}$$

which, for $A_d = -100$, gives $R_C = 5.4$ kΩ (less than 7.5 kΩ).

Small-Signal Parameters

$R_o = R_C \| r_{02} = 5.4$ k $\|$ 113 k $= 5.15$ kΩ.
$r_\pi = \beta_{FN} V_T/I_C = 102 \times 25.8$ mV/1 mA $= 2632$ Ω.
$r_e = r_\pi/(1 + \beta_{FN}) = 25.6$ Ω.   (equivalent emitter resistance)
$R_{id} = 2 r_\pi = 2 \times 2632 = 5264$ Ω.

$$A_c = \frac{-(g_m/2)(R_C \| r_{02})}{1 + 2g_m R_{EE}(1 + 1/\beta_F)} = \frac{-(38.76 \text{ mA}/2) \times (5.4 \text{ k} \| 113 \text{ k})}{1 + 2 \times 38.7 \text{ mA/V} \times 56.5 \text{ k}(1 + 1/102)} = -0.0228$$

CMRR $\approx 2 g_m R_{EE} = 2 \times 38.7$ mA/V $\times 56.5$ k $= 4379$ (or 72.8 dB).

$R_{ic} = r_\pi + (1 + \beta_F) 2 R_{EE} = 2632 + (1 + 102) \times 2 \times 56.5$ k $= 11.64$ MΩ.

9.5 EWB SIMULATION

The EWB has many circuit models for many types of BJTs and FETs. First, you get the schematic of a BJT from the list of Active devices, or an FET from the list of FET devices. Then, you double-click on the device (BJT or JFET) and the library of BJTs (or FETs) will open. The library has models of many commercially available BJTs and FETs. Also, you can change the model parameters by choosing **Edit**.

We will use EWB to simulate the differential amplifier shown in Fig. 9.2 to verify the design specifications.

DC Biasing

The steps to follow are:

1. Open file FIG9_2.CA4 from the EWB file menu. Run the simulation.
2. Check that the input signal is 0 V, 1 kHz. Otherwise, change the settings.
3. Check that the meters are set to DC. Double-click on the meter and then change, if needed.
4. Complete Table 9.1.

Differential Small-Signal

The steps to follow are:

1. Open file FIG9_3.CA4 from the EWB file menu. Run the simulation.

2. Check that the input signal is 10 mV, 1 kHz. Otherwise, change the settings.

3. Check that the meters are set to AC, except for I_Q and I_{ref}. Double-click on the meter and then change, if needed.

Table 9.1

	Calculated	Simulated	Practically Measured
I_{C1} (mA)	1	1.08	
I_{C2} (mA)	1	1.08	
I_B (µA)	9.8	1.06	
I_Q (mA)	2	2.17	
I_{ref} (mA)	2	1.98	
V_{o1} (V)	9.6	9.16	
V_{o2} (V)	9.6	9.16	
$V_{o(diff)}$ (V)	0	-2.22 µV	
V_E (V)	-0.7	-0.698	
V_{CE1} (V)	10.3	9.86	
V_{CE2} (V)	10.3	9.86	

4. Check that the Oscilloscope has settings of AC; time base: 0.1 ms/div and Y/T; channel A: 5 mV/div; channel B: 500 m/div. Otherwise, change the settings to give a clear display. *If the simulation does not reach the steady-state condtion, stop the simulation after 20 ms by clickimg the stop button..*

5. Zoom the oscilloscope display. Use the cursor to read the v_{o2} and v_d. Find the voltage gain.

$V_{o2(peak)}$ _____ (1.454 V) $V_{d(peak)}$ _____ (0.01414 V) $A_d = V_{o2(peak)}/V_{d(peak)}$ _____ (102.8)

6. Compare the input and output waveforms. Do you see crossover distortion? _____ (No).

7. Increase the value of the input voltage, and carefully observe the resulting output voltage waveform. What are the findings?

_____

_____

_____

What are the maximum and minimum values of the sinusoidal output voltages without any clipping?

$V_{o2(max)}$ _____ $V_{id(max)}$ _____ ($\approx$ 50 mV) $V_{o2(min)}$ _____ $V_{id(min)}$ _____ ($\approx$ -50 mV)

8. Complete Table 9.2.

Table 9.2

	Calculated	Simulated	Practically Measured
i_b (μA)	--	1.97	
v_{id} (V)	--	10	
v_{o1} (V)	--	1.03 V	
v_{o2} (V)	--	1.03 V	
$v_{o(diff)}$ (V)	--	2.06 V?	
v_E (mV)	0	5.01	
$A_d = v_{o2}/v_{id}$	100	1.03 V/10 mV = 103	
$A_d = v_{o(dif)}/v_{id}$	200	2.06 V/10 mV = 206	
$R_{id} = v_{id}/i_b$ (Ω)	5264	10 mV/1.9 μV = 5263	
R_o (kΩ)	5.15	5.169	

9. To find the output resistance R_o. Run the simulation for $R_L = 10$ kΩ.

$$R_o = R_L(=10\text{ k}\Omega)\left[\frac{v_{o2}(\text{for } R_L = 1000\,M\Omega)}{v_{o2}(\text{for } R_L = 10\text{ k}\Omega)} - 1\right] \quad (9\text{-}13)$$

R_o, $= 10$ kΩ. x $(1.03$ V/679 mV $- 1) = 5.169$ kΩ.

10. (Only for EWB version 5.0) Run the worst-case DC analysis for the minimum and maximum output voltages of the amplifier in Fig. 9.4 at node 6 (collector of transistor Q_2). Set tolerance = 20%, output node = 6, and Min/Max voltage. Plot the minimum and maximum output voltages and discuss the effects of tolerance.

_____

_____

_____

11. (Only for EWB version 5.0) Run the Monte-Carlo transient analysis for the output voltage of the amplifier in Fig. 9.4 at node 6 (collector of transistor Q_2). Set the dialog box as shown in Fig. 9.5: number of runs = 3, tolerance = 20%, distribution = uniform, and transient analysis. Plot the variations of output voltage with the number of runs. Comment on the results.

_____

_____

_____

Figure 9.5
Setting Monte-Carlo analysis

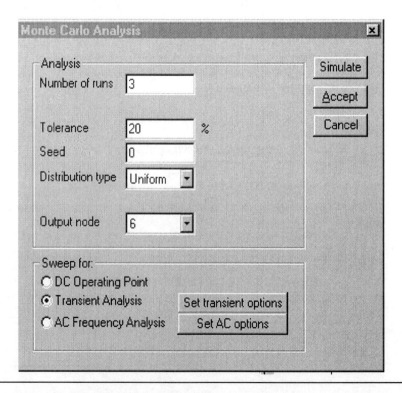

Common Mode Small-Signal

The steps to follow are:
1. Open file FIG9_4.CA4 from the EWB file menu. Run the simulation.
2. Check that the input signal is 4 V, 1 kHz. Otherwise, change the settings.
3. Check that the meters are set to AC, except for I_Q and I_{ref}. Double-click on the meter, and then change, if needed.
4. Check that the Oscilloscope has setting of AC; time base: 0.1 ms/div and Y/T; channel A: 2 V/div; channel B: 100 m/div. Otherwise, change the settings to give a clear display.
5. Zoom the oscilloscope display. Use the cursor to read the v_{o2} and v_{ic}. Find the voltage gain.

$V_{o2(peak)}$ _____ (0.25916 V) $V_{ic(peak)}$ _____ (5.6586 V) $A_d = V_{o2(peak)}/V_{ic(peak)}$

_____ (0.0458)

6. Compare the input and output waveforms. What are the findings? (180° phase shift)

_____
_____
_____

7. Complete Table 9.3.

Table 9.3

	Calculated	Simulated	Practically Measured
i_b (μA)	--	0.69	
V_{ic} (V)	--	4 V (rms)	
V_{o1} (mV)	--	183	
V_{o2} (mV)	--	183	
$V_{o(diff)}$ (mV)	--	92.8	
V_E (V)	0	4.0	
$A_c = v_{o2}/v_{ic}$	0.0228	183 mV/4 V = 0.0458	
$A_{c(diff)} = v_{o(diff)}/v_{ic}$	0.0456	92.8 mV/4 V = 0.0232	
$R_{ic} = v_{ic}/i_b$ (MΩ)	11.64	4 V/0.69 μA = 5.8 MΩ	

8. Increase the value of the input voltage and carefully observe the resulting output voltage waveform. Observe any change or distortion of the output waveform (which is expected to be identical to the sinusoidal input signal). What are the maximum and minimum values of the sinusoidal output voltages at which the clipping or distortion occurs?

$V_{o2(max)}$ _____ (0.427 V) $V_{ic(max)}$ _____ (9.34 V) $V_{o2(min)}$ _____ (0.427 V) $V_{ic(min)}$ _____ (-9.34 V)

9.6 WRITE-UP/CONCLUSIONS

Write a brief report, summarizing the results of the design experiment and what you have learned or confirmed about the differential amplifiers.

Comment on the relationship between theory, simulation and practical circuit performance and the precautions that must be taken when using this learning approach.

_____
_____
_____
_____
_____
_____

9.7 REINFORCEMENT EXERCISES

Two resistances R_{e1} and R_{e2}, each of 50 Ω are connected to the emitters of transistors Q_1 and Q_2 as shown in Fig. 9.5. The differential voltage gain can be found approximately from

$$A_d = \frac{R_{C2}}{r_{e2} + R_{e2}} = \frac{R_C}{r_e + R_e} \qquad (9\text{-}14)$$

where $r_e = r_\pi/(1 + \beta_F)$ is the emitter resistance of a bipolar transistor.

Figure 9.5
BJT differential amplifier with emitter resistors

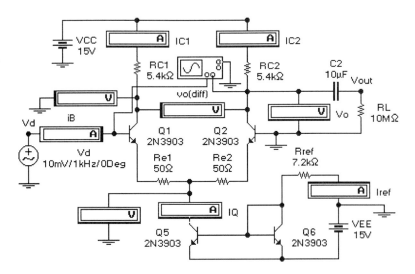

1. Open file FIG9_5.CA4 from the EWB file menu. Run the simulation.
2. Check that the input signal is 10 mV, 1 kHz. Otherwise, change the settings.
3. Check that the meters are set to AC, except for I_Q and I_{ref}. Double-click on the meter and then change, if needed.
4. Check that the Oscilloscope has settings of AC; time base: 0.1 ms/div and Y/T; channel A: 5 mV/div; channel B: 200 m/div. Otherwise, change the settings to give a clear display. *If you don't see the Oscilloscope screen, click the right mouse button on the Oscilloscope to open it.*
5. Complete Table 9.4.
6. Compare the results of Table 9.4 with those of Table 9.3. What are the effects of emitter resistances on the differential gain A_d and input resistance R_{id}? Explain.

_____
_____
_____

Table 9.4

	Calculated	Simulated	Practically Measured
i_b (μA)	--	0.67	
V_{id} (mV)	--	10	
V_{o1} (mV)	--	350	
V_{o2} (mV)	--	350	
$V_{o(diff)}$ (mV)	--	700	
V_E (mV)	--	5.0	
$A_d = v_{o2}/v_{id}$	71.4	35.0	
$R_{id} = v_{id}/i_b$ (kΩ)	15.56	14.93	

9.8 DESIGN PROBLEMS

There may be more than one solution to the following problems. Use EWB or PSPICE to verify your design. Also, build and test, if possible. Determine the voltage and current ratings of active and passive components.

1. Design the BJT differential amplifier, shown in Fig. 9.2, to give a differential mode voltage gain, $|A_d| = |v_{od}/v_{id}| \geq 500$ at a load resistance of $R_L = 50$ kΩ. Assume the source resistance is $R_s = 0$ Ω, and the DC supply voltages are $V_{CC} = -V_{EE} = 15$ V.

2. Design the BJT differential amplifier, shown in Fig. 9.4, to give a common-mode rejection ratio of CMRR $\geq 5,000$ at a load resistance of $R_L = 50$ kΩ. Assume the source resistance is $R_s = 0$ Ω, and the DC supply voltages are $V_{CC} = -V_{EE} = 15$ V.

3. Design an FET differential amplifier to give a differential mode voltage gain, $|A_d| = |v_{od}/v_{id}| \geq 250$ at a load resistance of $R_L = 50$ kΩ. Assume the source resistance is $R_s = 0$ Ω, and the DC supply voltages are $V_{DD} = -V_{SS} = 15$ V.

10 DESIGN OF OPERATIONAL AMPLIFIERS

10.1 LEARNING OBJECTIVES

To design operational amplifier (op-amp) circuits to perform a specified function with a specified voltage gain. We will use Electronics Workbench to verify the design and evaluate the performance of the op-amp circuits.

At the end of this lab, you will
- Be familiar with the characteristics and basic applications of op-amps to perform specified wave-shaping functions.
- Be able to analyze and design simple op-amp circuits to meet certain specifications.

10.2 THEORY

An operational amplifier (or op-amp) is a high gain, direct-coupled differential amplifier that has a high input resistance (i.e., from 2 MΩ to 10^{12} Ω), a high voltage gain (i.e., from 10^4 to 10^6) and a low output resistance (i.e., from 50 to 75 Ω). An op-amp serves as a building block for many electronic circuits. For most applications, one can design op-amp circuits from the knowledge of terminal characteristics of op-amps. The symbol of an op-amp is as shown. It has at least five terminals as follows:

1. Terminal v- is called the *inverting input* because the output due to an input at this terminal will be inverted.
2. Terminal v+ is called the *non-inverting input* because the output due to an input at this terminal will have the same polarity.
3. Terminal - is for negative DC supply -V_{EE}.
4. Terminal v_o is the *output* terminal.
5. Terminal + is for positive DC supply +V_{CC}.

An op-amp operates with a differential voltage between the two input terminals. The output voltage is directly proportional to the small-signal differential (or difference) input voltage. That is, the output voltage v_o is given by

$$v_o = A_o v_d = A_o (v_+ - v_-) \tag{10-1}$$

where A_o = open-loop voltage gain, v_d = small-signal differential (or difference) input voltage, v_- is the small-signal voltage at the inverting terminal with respect to the ground, and v_+ is the small-signal voltage at the non-inverting terminal with respect to the ground.

The input current drawn by the amplifier is very small (typically of the order of nA), tending to be zero. The analysis and design of op-amp circuits can be greatly simplified, if an op-amp is assumed to be an "ideal" one, which is characterized by

1. The open-loop voltage gain is infinite, $A_o = \infty$.
2. The input resistance is infinite, $R_i = \infty$.
3. The amplifier draws no current, $i_i = 0$.
4. The output resistance is negligible, $R_o = 0\,\Omega$.
5. The gain A_o remains constant, and it is not a function of frequency.

The output voltage v_o will depend on the three possible conditions: (a) $v_- = 0$ in which case v_o will be positive (i.e., $v_o = A_o v_+$), (b) $v_+ = 0$ in which case v_o will be negative (i.e., $v_o = -A_o v_-$), and (c) both inputs v_+ and v_- are present such that $v_o = A_o (v_+ - v_-)$. Therefore, depending upon the conditions of the input voltages, op-amp circuits can be classified into three basic configurations:

Non-Inverting Amplifier
Inverting Amplifier
Differential (or Difference) Amplifier

Non-Inverting Amplifier

The configuration of a non-inverting amplifier is shown in Fig. 10.1. The input voltage v_s is connected to the non-inverting terminal. R_F provides feedback from the output to the input side of the op-amp. $R_x = (R_1 || R_F)$ is used to minimize the effect of offset biasing current on the output. The *closed-loop voltage gain* A_f can be found from

$$A_f = \frac{v_o}{v_s} = 1 + \frac{R_F}{R_1} \qquad (10\text{-}2)$$

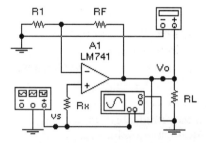

Figure 10.1
Non-inverting amplifier

The input resistance R_{in} and the output resistance R_{out} of the amplifier are given by

$$R_{in} = \frac{v_s}{i_s} = \infty \qquad (10\text{-}3)$$

$$R_{out} = R_o \approx 0\,\Omega \qquad (10\text{-}4)$$

Inverting Amplifier

Another common configuration is the inverting voltage amplifier as shown in Fig. 10.2. R_F is used to feedback the output voltage to the inverting terminal of the op-amp. $R_x = (R_1 || R_F)$. The *closed-loop voltage gain* can be found from

$$A_f = \frac{v_O}{v_s} = -\frac{R_F}{R_1} \qquad (10\text{-}5)$$

The input resistance R_{in} and the output resistance R_{out} of the amplifier are given by

$$R_{in} = \frac{v_s}{i_s} = R_1 \qquad (10\text{-}6)$$

$$R_{out} = R_o \approx 0\,\Omega \qquad (10\text{-}7)$$

Figure 10.2
Inverting amplifier

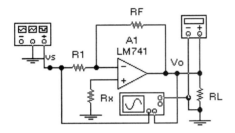

Differential Amplifier

The differential (or difference) amplifier is shown in Fig. 10.3. Two input voltages (v_s, v_a) are applied, one to the non-inverting terminal and another to the inverting terminal. Resistances R_a and R_x are used to step-down the voltage applied to the non-inverting terminal. $R_x = (R_1 || R_F)$. Using the superposition theorem to find the output voltage v_{os} due to v_s and the output voltage v_{oa} due to v_a, the resultant output voltage v_o is given by

$$v_o = v_{os} + v_{oa} = -\frac{R_F}{R_1}v_s + \left(1 + \frac{R_F}{R_1}\right)\left(\frac{R_x}{R_x + R_a}\right)v_a \qquad (10\text{-}8)$$

which, for $R_a = R_1$ and $R_F = R_x$, becomes

$$v_o = \frac{R_F}{R_1}(v_a - v_s) \qquad (10\text{-}9)$$

Thus, the circuit can operate as a difference voltage amplifier with a closed-loop voltage gain of R_F/R_1.

Figure 10.3
Difference amplifier

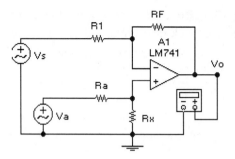

10.3 READING ASSIGNMENT

Study chapter(s) on op-amp circuits.

10.4 ASSIGNMENT

Design Specifications

We will design the non-inverting, inverting, and difference amplifiers to give a specified voltage gain. Also, we will extend the basic circuits to perform summation, integration, and differentiation. Use op-amp LM741 whose $R_i = 2$ MΩ, $A_o = 10^5$, and $R_o = 75$ Ω. Assume DC supply voltages, $V_{CC} = -V_{EE} = 21$ V.

Non-Inverting Amplifier

Design the non-inverting amplifier shown in Fig. 10.1 to meet the following specifications:

- Closed-loop voltage gain, $A_f = v_o/v_s = 100$.
- Input signal voltage, $v_s = 10$ mV, 1 kHz.

RESISTOR VALUES

Let us choose $R_1 = 10$ kΩ.
$A_f = 1 + R_F/R_1$, which, for $A_f = 100$, gives $R_F = 990$ kΩ.

Inverting Amplifier

Design the inverting amplifier shown in Fig. 10.2 to meet the following specifications:

- Closed-loop voltage gain, $A_f = v_o/v_s = -100$.
- Input signal voltage, $v_s = 10$ mV, 1 kHz.
- The input current from the signal source, $i_s \leq 1$ μA.

RESISTOR VALUES

$R_{in} \geq v_s/i_s = 10$ mV/1 μA $= 10$ kΩ. So, let us choose $R_1 = 10$ kΩ.
$A_f = -R_F/R_1$ which, for $A_f = -100$, gives $R_F = 1$ MΩ.

Differential (Difference) Amplifier

Design the differential amplifier shown in Fig. 10.3 to meet the following specifications:

- The input signal voltages, $v_s = 1$ mV, 1 kHz, and $v_a = 1.5$ mV, 1 kHz.
- The output voltage gain, $v_o = 100$ ($v_a - v_s$).
- The input current from the signal source v_s must be less than 1 μA.

RESISTOR VALUES

$R_{in} \geq v_s/i_s = 10$ mV/1 μA $= 10$ kΩ, so let us choose $R_1 = 10$ kΩ.

Inverting Summing Amplifier

$A_f = R_F/R_1$ which, for $A_f = 100$, gives $R_F = 10$ MΩ.

The inverting amplifier in Fig. 10.2 can be extended to sum two or more voltage signals by adding a resistance for each input signal. This is shown in Fig. 10.4 for two input signals, v_{s1} and v_{s2}. The output voltage is given by

$$v_o = -\frac{R_F}{R_1}v_{s1} - \frac{R_F}{R_2}v_{s2} \tag{10-10}$$

which, for $R_1 = R_2$, becomes

$$v_o = -\frac{R_F}{R_1}(v_{s1} + v_{s2}) \tag{10-11}$$

Figure 10.4
Inverting summing amplifier

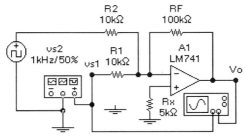

Design the summing amplifier shown in Fig. 10.4 to meet the following specifications:
- The input signal voltages $v_{s1} = 1$V, 1 kHz (sine wave), and $v_{s2} = 1$ V, 1 kHz (square wave).
- The output voltage gain, $v_o = -10\,(v_{s1} + v_{s2})$.
- The input current from any source must be less that 100 µA.

RESISTOR VALUES

$R_{in} \geq v_s/i_s = 1$ V/100 µA $= 10$ kΩ, so let us choose $R_1 = 10$ kΩ.
$A_f = -R_F/R_1$ which, for $A_f = 10$, gives $R_F = 100$ kΩ.

Inverting Integrator

If the resistance R_F in the inverting amplifier of Fig. 10.2 is replaced by a capacitor C_F, the circuit will operate as an integrator. This is shown in Fig. 10.5. The impedance of C_F in Laplace's domain is $Z_F = 1/(sC_F)$. Thus, the output voltage in Laplace's domain becomes

$$V_o(s) = -\left(\frac{Z_F}{Z_1}\right)V_s(s) = -\frac{1}{sR_1C_F}V_i(s) \tag{10-12}$$

which gives the output voltage in the time domain as

$$v_o(t) = -\frac{1}{R_1C_F}\int_0^t v_s\,dt + v_c(t=0) \tag{10-13}$$

where $v_c(t=0) = V_{co}$ represents the initial capacitor voltage. Thus, the output voltage is the negative integral of the input voltage v_s. A large value resistor R_F

is, normally, connected in parallel with the capacitor C_F. R_F provides the DC feedback and overcomes the saturation problem. Time constant τ_F (= $R_F\,C_F$) must be larger than the period $T(=1/f_s)$ of the input signal. A ratio of ten to one is, generally, adequate such that $\tau_F = 10\,T$.

Figure 10.5
Inverting integrator

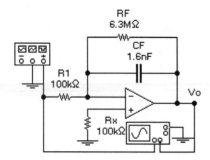

Design the inverting integrator shown in Fig. 10.5 to meet the following specifications:
- The input signal voltage v_s = 1V, 1kHz (sine wave or square wave).
- The output voltage v_o = -10 $\int v_s\,dt$ (with zero initial conditions).
- The input current from the source must be less than 10 µA.

RESISTOR AND CAPACITOR VALUES

$T = 1/f_s = 1$ ms.
$\omega_s = 2\,\pi\,f_s = 2\,\pi \times 1$ kHz = 2π krad/s.
$R_{in} \geq v_s/i_s = 1V/10\,\mu A = 100\,k\Omega$.
Thus, let us choose $R_1 = 100\,k\Omega$.
$A_f = -1/(\omega_s\,R_1\,C_F)$, which, for $A_f = -10$, gives $C_F = 1.6$ nF.
$R_F\,C_F = 10\,T = 10$ ms, which, for $C_F = 1.6$ nF, gives $R_F = 6.3\,M\Omega$.

Inverting Differentiator

If the resistance R_1 in the inverting amplifier of Fig. 10.2 is replaced by a capacitor C_1 as shown in Fig. 10.6, the circuit will operate as a differentiator. The impedance of C_1 in Laplace's domain, $Z_1 = 1/(sC_1)$. The output voltage in Laplace's domain is

$$V_o(s) = -\left(\frac{R_F}{Z_1}\right)V_s(s) = -sR_F C_1 V_s(s) \qquad (10\text{-}14)$$

which gives the output voltage in the time domain as

$$v_o(t) = -R_F C_1 \frac{dv_s}{dt} \qquad (10\text{-}15)$$

In order to avoid the output due to any sharp change in the input voltage $v_s(t)$ such as noise or 'picked-up' interference, a resistor R_1 (< R_F) is connected in series with C_1 in order to limit the gain at high frequencies.

Figure 10.6
Inverting differentiator

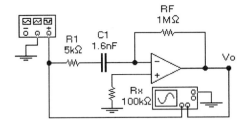

Design the inverting differentiator shown in Fig. 10.6 to meet the following specifications:
- The input signal voltage $v_s = 1$ V, 1kHz (sine wave or triangular wave).
- The output voltage $v_o = -10\, dv_s/dt$.
- The frequency for the differentiation to be ineffective, $f_b = 5$ kHz.

RESISTORS AND CAPACITOR VALUES

$T = 1/f_s = 1$ ms. $\omega_s = 2\pi f_s = 2\pi \times 1$ kHz $= 2\pi$ krad/s, so, let us choose a small value of C_1, say $C_1 = 1.6$ nF. There is no unique value for C_1, but as a general guideline, the lower the value, the less expensive it is.

$A_f = -\omega_s R_1 C_1$, which, for $A_f = -10$, gives $R_F = 1$ MΩ.
$\omega_b = 2\pi f_b = 2\pi \times 5$ kHz $= 10\pi$ krad/s.
$\omega_b = 2\pi f_b = 1/R_1 C_1$, which gives $R_1 = 5$ kΩ.

10.5 EWB SIMULATION

The EWB has many circuit models for many types of op-amps. First, you get the schematic of an op-amp from the list of Active devices. Then, you double-click on the op-amp and the library of op-amps will open. The library has models of many commercially available op-amps. Also, you can change the model parameters by choosing **Edit**.

We will use EWB to simulate the op-amp circuits to verify the design specifications.

Non-Inverting Amplifier

The steps to follow are:
1. Open file FIG10_1.CA4 from the EWB file menu.
2. Check that the input signal is 10 mV at 1 kHz. Otherwise, change the settings.
3. Check that the multimeter is set to AC. Run the simulation.
4. Check that the Oscilloscope has settings of AC; time base: 0.2 ms/div (comes up as 0.2 ms/div) and Y/T; channel A: 5 mV/div; channel B: 500 m/div. Otherwise, change the settings to give a clear display.
5. Zoom the oscilloscope display. Use the cursor to read the v_o and v_s. Find the voltage gain.
$V_{o(peak)}$ _____ (0.99885 V) $V_{s(peak)}$ _____ (0.010003 V) $A_v = V_{o(peak)}/V_{s(peak)}$ _____ (99.86)
6. Complete Table 10.1.

115

7. To find the output resistance R_o, run the simulation for R_L = 10 kΩ, 10 MΩ.

$$R_o = R_L(=10\text{ k}\Omega)\left[\frac{v_o(\text{for }R_L=10\text{ M}\Omega)}{v_o(\text{for }R_L=10\text{ k}\Omega)}-1\right] \quad (10\text{-}16)$$

$R_{o,}$ = 10 kΩ. x (0.705 V/0.705 V - 1) = 0 Ω.

Table 10.1

	Calculated	Simulated	Practically Measured
i_s	--	0 μA	
v_o	--	705 mV	
$A_v = v_o/v_s$ (rms)*	100	705/(10/√2) = 70.5 x √2 = 99.7	
R_o (Ω)	0	0.0	
$R_{in} = v_s/i_s$ (kΩ)	∞	(10 mV/√2)/0 μA = ∞	

Note: Function generator 'amplitude' is zero to peak value. For sinusoidal output $V_{s(rms)} = V_{s(peak)}/\sqrt{2}$.

Inverting Amplifier

The steps to follow are:
1. Open file FIG10_2.CA4 from the EWB file menu.
2. Check that the input signal is 10 mV, 1 kHz. Otherwise, change the settings.
3. Check that the multimeter is set to AC. Run the simulation.
4. Check that the Oscilloscope has settings of AC; time base: 0.1 ms/div (comes up as 0.2 ms/div) and Y/T; channel A: 10mV/div; channel B: 1 V/div. Otherwise, change the settings to give a clear display.
5. Zoom the oscilloscope display. Use the cursor to read the v_o and v_s. Find the voltage gain.

$V_{o(peak)}$ _____ (-1.9976 V) $V_{s(peak)}$ _____ (2.0006 x 10^{-2} V) A_v = $V_{o(peak)}/V_{s(peak)}$ _____ (-99.85).

6. To find the output resistance R_o, run the simulation for R_L = 10 kΩ, 10 MΩ.

$$R_o = R_L(=10\text{ k}\Omega)\left[\frac{v_o(\text{for }R_L = 10\text{ M}\Omega)}{v_o(\text{for }R_L = 10\text{ k}\Omega)} - 1\right] \quad (10\text{-}17)$$

$$R_o, = 0\ \Omega.$$

7. Complete Table 10.2.

Table 10.2

	Calculated	Simulated	Practically Measured
i_s	--	1.41 μA	
v_o	--	1.41 V	
$A_v = v_o/v_s$ (rms)*	-100	1.41 V/(2 x 10 mV/√2) = 99.7	
R_o (Ω)	0	0	
$R_{in} = v_s/i_s$ (kΩ)	10	2 x 10 mV/ (√2 x 1.41 μA) = 10.03 kΩ	

* *Note*: If we use v_o = 1.41 V and v_s = 10 mV, we can calcuate the voltage gain as A_v= 1.41 V/10 mV = 141 or A_v= 1.41 V/(10 mV/√2) = 199.4. But, both of thes numbers are not correct! The EWB's Function Generator can be very confusing if we are not careful about the - to + connection versus common to + connection.. We will never get the negative (-) sign on the voltage gain from the meters, but we can get only by looking on the oscilloscope.

8. Compare the results in Table 10.2 with those in Table 10.1. Both the inverting and non-inverting amplifiers can give the desired voltage gain. In what ways do they differ? Explain.

_____

_____

_____

Difference Amplifier

The steps to follow are:
1. Open file FIG10_3.CA4 from the EWB file menu.
2. Check that the input signals are $v_s = 1$ mV, 1 Hz and $v_a = 1.5$ mV, 1 Hz. Otherwise, change the settings.
3. Check that the multimeter is set to AC. Run the simulation.
4. Complete the values in Table 10.3 for f = 1 kHz.
5. Change the frequency of the two input signals to 10 kHz, and then complete the values in Table 10.3 for f = 10 kHz.
6. You will notice that the gain has decreased with the frequency. Why? Explain.

_____

_____

_____

Table 10.3

	Calculated	Simulated	Practically Measured
$v_a - v_s$ (mV)	0.5	0.5 mV	
v_o at 1 Hz	--	50 mV	
v_o at 10 kHz	--	41.3 mV	
$A_v = v_o/(v_a - v_s)$ at 1 Hz	100	100	
$A_v = v_o/(v_a - v_s)$ at 10 kHz	100	82.6	

Inverting Summing Amplifier

The steps to follow are:
1. Open file FIG10_4.CA4 from the EWB file menu. Run the simulation.

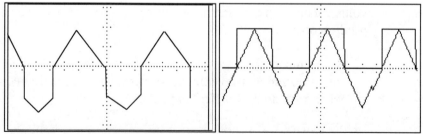

Note: The jogged shape on the output is probably caused during importing the graphic to this manuscript. You may not see it the on the simulation screen

2. Check that the input signal v_{s1} is 1 V, 1 kHz (triangular wave), and v_{s2} is 1 V, 1 kHz (square wave). Otherwise, change the settings.
3. Plot the expected output waveform beside the input waveforms as shown.
4. Check that the Oscilloscope has settings of AC; time base: 0.2 ms/div and Y/T; channel A: 500 mV/div; channel B: 5 m/div. Otherwise, change the settings to give a clear display. *If the simulation does not reach the steady-state condtion, stop the simulation by clicking the stop button.*
5. Compare the output waveform with the expected waveform. Comment on any differences.

_____

_____

_____

Inverting Integrator

The steps to follow are:
1. Open file FIG10_5.CA4 from the EWB file menu. Run the simulation.
2. Check that the input signal is a square wave 1 V, 1 kHz. Otherwise, change the settings.
3. Plot the expected output waveform on top of the input waveform.
4. Check that the Oscilloscope has settings of AC; time base: 0.2 ms/div and Y/T; channel A: 1V/div; channel B: 1V/div. Otherwise, change the settings to give a clear display.
5. Compare the output waveform with the expected waveform. Comment on any differences.

_____

_____

_____

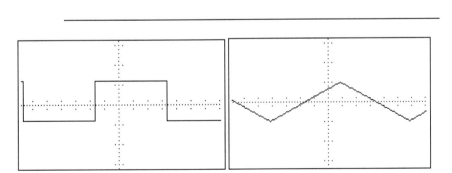

6. Change the input signal to a triangular wave of 1 V, 1 kHz by clicking on the function generator.
7. Plot the expected output waveform on the top of the input waveform.

8. Compare the output waveform with the expected waveform. Comment on any differences.

_____
_____
_____

Inverting Differentiator

The steps to follow are:
1. Open file FIG10_6.CA4 from the EWB file menu. *Check that the channel A is connected to the input side and the channel B is connected to the outpuit side. Otherwise, make the connections.* Run the simulation.
2. Check that the input signal is a square wave 1 V, 1 kHz. Otherwise, change the settings.
3. Plot the expected output waveform on top of the input waveform.
4. Check that the Oscilloscope has settings of AC; time base: 0.2 ms/div and Y/T; channel A: 1V/div; channel B: 1V/div. Otherwise, change the settings to give a clear display.
5. Compare the output waveform with the expected waveform. Comment on any differences. Then, change $R_F = 100$ kΩ, $R_1 = 10$ kΩ and run the simulation. Comment on the effect of the time consatnt on the output waveform.

_____
_____
_____

6. Change the input signal to a triangular wave 1 V, 1 kHz.
7. Plot the expected output waveform on top of the input waveform.
8. Check that the Oscilloscope has settings of AC; time base: 0.2 ms/div and Y/T; channel A: 1V/div; channel B: 1V/div. Otherwise, change the settings to give a good readable display.

9. Compare the output waveform with the expected waveform. Comment on any differences.

_____
_____
_____

10.6   WRITE-UP/CONCLUSIONS

Write a brief report, summarizing the results of the design experiment and what you have learned or confirmed about the op-amp circuits.

Comment on the relationship between theory, simulation and practical circuit performance and the precautions that must be taken when using this learning approach.

_____
_____
_____
_____
_____
_____
_____

10.7   REINFORCEMENT EXERCISES

1. This test will reveal one of the limitations of the op-amp itself.
A. Open file FIG10_7.CA4 from the EWB file menu. This is the same circuit as the non-inverting amplifier in Fig. 10.1.
B. Check that the input signal is 5V (sine wave) at 1 MHz. Otherwise, change the settings.
C. Check that the multimeter is set to AC. Run the simulation.
D. Check that the Oscilloscope has settings of AC; time base: 0.1 μs/div and Y/T; channel A: 5 V/div; channel B: 100 mV /div. Otherwise, change the settings to give a clear display. *If the 'auto' trigger does not work, change to channel A triggering.*

E. Compare the output with the input. Is the output a sine wave? Why? What causes this distortion? Explain.

_____

_____

_____

Find the slope v_s/dt: _____ (0.2436 V/0.4948 μs = 0.492 V/μs). Compare it with the op-amp slew rate: 0.5 V/μs.
F. Change the input signal to 5V (square wave) at 1 MHz.
G.
H. Compare the output with the input. Is the output still a square wave? Why? Explain.

_____

_____

_____

Find the slope v_s/dt: _____ (0.24811 V/0.4958 μs = 0.5004 V/μs). Compare it with the op-amp slew rate: 0.5 V/μs.

2. The difference amplifier is shown in Fig. 10.7. v_s is square wave of 1 V at 1 kHz, and v_s is a triangular wave of 1 V at 1 kHz.
The steps to follow are:
A. Open file FIG10_8.CA4 from the EWB file menu. *If the setting on the oscilloscope channel B is too small (at 1 V/div), change it to 2V/div).* Run the simulation.
B. Plot the expected output waveform on the top of the input waveforms.
C. Compare the output waveform with the expected waveform. Comment on any differences.

_____

_____

_____

Figure 10.8
Difference amplifier

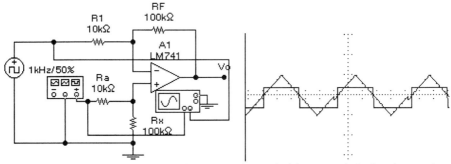

Note: The jogged shape on the output is probably caused during importing the graphic to this manuscript. You may not see it the on the simulation screen

10.8 DESIGN PROBLEMS

There may be more than one solution to the following problems. Use EWB or PSPICE to verify your design. Also, build and test, if possible. Determine the voltage and current ratings of active and passive components.

1. Design an op-amp integrator which is to be operated with an AC signal of 5 kHz and should give a voltage gain of -10 at a specified low frequency of $\omega = 10$ rad/s. The input current from the source must be less than 1 µA. The DC supply voltages are $V_{CC} = -V_{EE} = 12$ V.

2. Design an op-amp differentiator which should give a maximum voltage gain of $A_{f(max)} = 20$ at the gain limiting frequency of $f_b = 10$ kHz. The input current from the source must be less than 1 µA. The DC supply voltages are $V_{CC} = -V_{EE} = 12$ V. Determine the voltage and current ratings of active and passive components.

3. Design an op-amp difference circuit to give an output voltage $v_o = 1000 (v_{s1} - v_{s2})$ where v_{s1} and v_{s2} are the input signals. The input current from the source must be less than 1 nA. The DC supply voltages are $V_{CC} = -V_{EE} = 15$ V. Determine the voltage and current ratings of active and passive components.

4. Design a precision full-wave rectifier (using op-amp and diodes) to convert an AC sinusoidal input signal of 100 µV at a frequency of 100 kHz to a DC voltage. The input current from the source must be less than 1 µA. The DC supply voltages are $V_{CC} = -V_{EE} = 12$ V. Determine the voltage and current ratings of active and passive components.

11 DESIGN OF AMPLIFIERS FOR FREQUENCY RESPONSE

11.1　LEARNING OBJECTIVES

To design common-emitter (BJT) and common-source (FET) amplifiers to operate in a specified frequency range. We will use Electronics Workbench to verify the design and evaluate the performance of the amplifier.

At the end of this lab, you will

- Be familiar with the frequency characteristics of common-emitter and common-source amplifiers, and the circuits elements that contribute to the frequency response.
- Be able to analyze and design common-emitter and common-source amplifiers to meet certain frequency specifications.

11.2　THEORY

The current gain of BJTs and the transconductance gain of FETs are frequency dependent, and their values decrease with the input signal frequency. The internal capacitances set the upper frequency limit of transistor amplifiers. BJT and FET amplifiers are often connected to the input signal source and the load resistor through coupling capacitors, which effectively block signals at low frequencies. Capacitors are also employed to bypass resistors effectively in order to increase the small-signal voltage gain. Thus, the bypass and coupling capacitors set the low frequency limit. BJT (or FET) amplifiers exhibit band-pass characteristics. However, op-amp amplifiers exhibit low-pass characteristics. Thus, the performance of amplifiers depend upon the input frequency, and the design specifications usually quote the voltage gain over a specified frequency range known as the *bandwidth*.

The determination of frequency characteristics requires finding the break frequency, and the mid-band gain. The approximate values of the low break frequency can be found by the *short-circuit method,* and those of the high break frequencies by the *zero-value method*. The short-circuit method assumes that only one capacitor contributes to the 3-dB break frequency and the other capacitors are effectively short-circuited. Thus, the low 3-dB frequency is determined from the effective contributions of all capacitors. That is,

$$f_L = \frac{1}{2\pi}\sum_{k=1}^{n}\frac{1}{\tau_{ck}} = \frac{1}{2\pi}\sum_{k=1}^{n}\frac{1}{C_k R_{ck}} \tag{11-1}$$

where τ_{Ck} is the time constant due to the kth capacitor only, and R_{Ck} is the Thevenin's equivalent resistance presented to C_k. The zero-value method finds the time constant τ_{Cm} for the mth capacitor by considering one capacitor at a time while setting other capacitors to 0 (or effectively open-circuiting them). Thus, the high 3-dB frequency is determined from the effective time constant of all capacitors. That is,

$$f_H = \frac{1}{2\pi\sum_{m=1}\tau_{Cm}} = \frac{1}{2\pi\sum_{m=1}C_m R_{Cm}} \tag{11-2}$$

where τ_{Cm} is the time constant due to mth capacitor only, and R_{Cm} is the Thevenin's equivalent resistance presented to C_m.

Frequency Response of Common-Emitter Amplifier

The configuration of a common-emitter amplifier is shown in Fig. 11.1. It has a band-pass frequency characteristic. Determination of the frequency response will require calculating (a) the DC biasing to find the small-signal parameters of the transistors, (b) the mid-frequency voltage gain $A_{v(mid)}$, (c) the low cut-off frequency f_L, and (d) the high cut-off frequency f_H. In Chapter 5, we described how to find the DC biasing and the mid-frequency gain.

Figure 11.1
Common-emitter amplifier

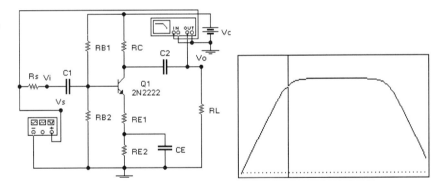

LOW BREAK FREQUENCIES

The low break frequencies depend upon the coupling-capacitors C_1 and C_2, and the bypass-capacitor C_E. The Thevenin's equivalent resistance presented to C_1 is

$$R_{C1} = R_s + R_B \| [r_\pi + (1+\beta_F)R_{E1}] \tag{11-3}$$

where $R_B = (R_{B1} \| R_{B2})$. The Thevenin's equivalent resistance presented to C_2 is given by

$$R_{C2} = R_C + R_L \tag{11-4}$$

The Thevenin's equivalent resistance presented to C_E is given by

$$R_{CE} = R_{E1} \left\| \frac{r_\pi + (R_s \| R_B)}{1 + \beta_F} \right.$$ (11-5)

If $R_s < R_B, r_\pi$; $R_{E1} > 1\ k\Omega$, and $\beta_F > 1$, Eq. (11-5) can be approximated by

$$R_{CE} \cong \frac{r_\pi}{\beta_F} = \frac{1}{g_m}$$ (11-6)

Therefore, the low cut-off frequency can be found from

$$f_L = \frac{1}{2\pi}\left[\frac{1}{C_1 R_{C1}} + \frac{1}{C_2 R_{C2}} + \frac{1}{C_E R_{CE}}\right]$$ (11-7)

HIGH BREAK FREQUENCIES

At high frequencies, the coupling and bypass capacitors offer very low impedances due to their high values, and can be assumed short-circuited. However, the internal capacitances (C_π and C_μ) of the transistor influence the high frequency breaks. C_π is the capacitance of the forward-biased base-emitter junction, and C_μ is the capacitance of the reverse-biased collector-base junction.

The resistance faced by C_π is given by

$$R_{C\pi} = r_\pi \| \left[(R_s \| R_B) + (1 + \beta_F)R_{E1}\right]$$ (11-8)

The Thevenin's equivalent resistance faced by C_μ is given by

$$R_{C\mu} = R_L \| R_C + R_i\left[1 + g_m(R_L \| R_C)\right]$$ (11-9)

where $R_i = (r_\pi \| R_B \| R_s)$. Thus, the high cut-off frequency f_H is given by

$$f_H = \frac{1}{2\pi(R_{C\pi}C_\pi + R_{C\mu}C_\mu)}$$ (11-10)

Common-Source Amplifier

The configuration of a common-source amplifier is shown in Fig. 11.2. Also, it has a band-pass frequency characteristic. In Chapter 6, we described how to find the DC biasing and the mid-frequency gain.

LOW BREAK FREQUENCIES

The low break frequencies depend upon the coupling-capacitors C_1 and C_2, and the bypass-capacitor C_S. The Thevenin's equivalent resistance presented to C_1 is

$$R_{C1} = R_s + R_G$$ (11-11)

where $R_G = (R_{G1} || R_{G2})$. The Thevenin's equivalent resistance presented to C_2 is given by,

$$R_{C2} = R_D + R_L \tag{11-12}$$

Figure 11.2
Common-source amplifier

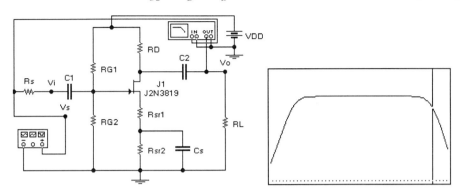

The Thevenin's equivalent resistance presented to C_S is given by

$$R_{CS} = R_{sr1} \left\| \frac{1}{g_m} \right. \tag{11-13}$$

Therefore, the low cut-off frequency can be found from

$$f_L = \frac{1}{2\pi} \left[\frac{1}{C_1 R_{C1}} + \frac{1}{C_2 R_{C2}} + \frac{1}{C_S R_{CS}} \right] \tag{11-14}$$

HIGH BREAK FREQUENCIES

At high frequencies, the coupling and bypass capacitors are assumed to be short-circuits. The Thevenin's equivalent resistance faced by C_{gs} is given by

$$R_{Cgs} = R_s \| R_G \tag{11-15}$$

The Thevenin's equivalent resistance faced by C_{gd} is given by

$$R_{Cgd} = (R_L \| R_D) + (R_S \| R_G)[1 + g_m (R_L \| R_D)] \tag{11-16}$$

Thus, the high cut-off frequency f_H is given by

$$f_H = \frac{1}{2\pi (R_{Cgs} C_{gs} + R_{Cgd} C_{gd})} \tag{11-17}$$

11.3 READING ASSIGNMENT

Study chapter(s) on frequency responses of amplifiers.

11.4 ASSIGNMENT

Design Specifications

We will design common-emitter and common-source amplifiers to give a specified mid-frequency voltage gain $A_{v(mid)}$, a low cut-off frequency f_L, and a high cut-off frequency f_H. Assume DC supply $V_{CC} = V_{DD} = 20$ V. In order to set the high cut-off frequency, a capacitor C_x is to be connected between the base and collector (or gate and drain) terminals of the transistor.

Common-Emitter Amplifier

Design the common-emitter amplifier shown in Fig. 11.1 to meet the following specifications:
- Midband voltage gain, $|A_{mid}| = v_o/v_s = 100$.
- Low cut-off frequency, $f_L = 1$ kHz.
- High cut-off frequency, $f_H = 250$ kHz
- Load resistance, $R_L = 10$ kΩ.
- Source resistance $R_s = 200$ Ω.
- Use NPN-transistor 2N2222 whose $\beta_F = 200$.
- Quiescent collector current, $I_C = 5$ mA (assumed).

RESISTOR VALUES

From Chapter, $R_C = 1.6$ kΩ, $R_{B1} = 68$ kΩ, $R_{B2} = 21$ kΩ, $R_E = 796$ Ω, $R_{E1} = 9$ Ω, and $R_{E2} = 787$ Ω, so

$r_\pi = \beta_F V_T/I_C = 200 \times 25.8$ mV/5 mA $= 1032$ Ω.
$g_m = \beta_F/r_\pi = 200/1032 = 193.8$ mA/V.
$R_B = R_{B1} || R_{B2} = 68$ k $|| 21$ k $= 16$ kΩ.

COUPLING AND BY-PASS CAPACITORS

$R_{C1} = R_s + R_B || [r_\pi + R_{E1}(1 + \beta_F)] = 2613$ Ω.

$R_{C2} = R_C + R_L = 11.6$ kΩ.

$$R_{CE} = R_{E1} \left\| \frac{r_\pi + (R_s \| R_B)}{1 + \beta_F} \right. = 14.83 \, \Omega$$

Set the break frequency corresponding to the lowest resistance as the low cut-off frequency f_L. That is,

$$f_{CE} = f_{L1} = f_L = 1 \text{ kHz}$$

$$f_{CE} = \frac{1}{2\pi R_{CE} C_E} = \frac{1}{2\pi \times 14.83 \times C_E} = 1 \text{ kHz or } C_E = 10.73 \, \mu F$$

Set the second break frequency corresponding to the next higher resistance. That is,

$$f_{L2} = f_{C1} = f_L/10 = 1 \text{ kHz}/10 = 100 \text{ Hz}$$

$$f_{C1} = \frac{1}{2\pi R_{C1} C_1} = \frac{1}{2\pi \times 2613\Omega \times C_1} = 100 \text{ Hz or } C_1 = 0.61 \ \mu F$$

Set the third break frequency corresponding to the highest resistance, $f_{L3} = f_{C2} = f_L/20 = 1 \text{ k}/20 = 50$ Hz.

$$f_{C2} = \frac{1}{2\pi R_{C2} C_2} = \frac{1}{2\pi \times 11.6 k\Omega \times C_2} = 50 \text{ Hz or } C_2 = 0.274 \ \mu F$$

ADDITIONAL SHUNT CAPACITOR

$V_{CE} = 8$ V, and $V_{BE} = 0.7$ V.
The forward base transit time, $\tau_F = 0.454$ ns.
The zero-biased base-emitter capacitance, $C_{jeo} = 19.5$ pF.
The zero-biased base-collector capacitance, $C_{\mu o} = 9.63$ pF.
$V_{CB} = V_{CE} + V_{EB} = V_{CE} - V_{BE} = 8 - 0.7 = 7.3$ V.
The base-charging capacitance, $C_b = \tau_F I_C/V_T = 0.454$ ns x 5 mA/25.8 mV = 87.98 pF.
The base-emitter junction capacitance can be found from

$$C_{be} = \frac{C_{beo}}{\sqrt{1 - V_{BE}/V_{be}}}$$

where V_{je} the base-emitter junction potential.
Assuming $V_{je} = 0.75$ V, we get
$C_{je} = 19.5 \text{ pF}/[1 - 0.7/0.75]^{0.5} = 75.52$ pF.
$C\pi = C_b + C_{je} = 87.98 \text{ pF} + 75.52 \text{ pF} = 163.5$ pF.
The collector-base junction capacitance can be found from

$$C_\mu = \frac{C_\mu}{\left(1 + V_{CB}/V_{jc}\right)^{0.5}}$$

where V_{je} the base-emitter junction potential.
Assuming $V_{jc} = 0.75$ V, we get
The collector-base junction capacitance can be found from
$C_\mu = 9.63 \text{ pF}/[1 + 7.3/0.75]^{0.5} = 2.94$ pF.
$R_{C\pi} = r_\pi || [(R_s || R_B + R_{E1}(1 + \beta_F)] = 681.7 \Omega$.
The Thevenin's equivalent resistance faced by C_μ is given by
$R_{C\mu} = (R_L || R_C) + (r_\pi || R_B || R_s)[1 + g_m(R_C || R_L)] = 45.86$ kΩ.
The high 3-dB frequency f_H is given by

$$f_H = \frac{1}{2\pi\left[R_{C\pi}C_\pi + R_{C\mu}(C_\mu + C_x)\right]} = 250 \text{ kHz or } C_\mu + C_x = 11.45 \text{ pF}$$

Common-Source Amplifier

which gives $C_x = 11.45 - 2.94 = 8.51$ pF. This is the additional capacitor C_x that is to be connected between the collector and base terminals of the BJT. For $C_x = 0$, $f_H = 646.22$ kHz.

Design the common-emitter amplifier shown in Fig. 11.2 to meet the following specifications:

- Midband voltage gain, $|A_{mid}| = v_o/v_s = 10$.
- Low cut-off frequency, $f_L = 1$ kHz.
- High cut-off frequency, $f_H = 250$ kHz.
- Load resistance, $R_L = 10$ kΩ.
- Source resistance, $R_s = 200$ Ω.
- Use JFET transistor J2N3919 whose $I_{DSS} = 11$ mA, $V_P = -2.5$ V, and $K_P = 1.76$ mA/V^2.
- Quiescent drain current, $I_D = 5$ mA (assumed).

RESISTOR VALUES

From Chapter 6, $R_{G1} = 1.7$ MΩ, $R_{G2} = 9$ MΩ, $R_D = 2.6$ kΩ, $R_{sr} = 798.4$ Ω, $R_{sr1} = 37.8$ Ω, $R_{sr2} = 760$ Ω.

$g_m = 5.93296$ mA/V.
$R_G = (R_{G1} || R_{G2}) = 1.43$ MΩ.

COUPLING AND BY-PASS CAPACITORS

$R_{C1} = R_s + R_G \equiv 1.43$ MΩ.
$R_{C2} = R_D + R_L = 12.6$ kΩ.
$R_{CS} = R_{sr} || (1/g_m) = 139 \Omega$.

Set the break frequency corresponding to the lowest resistance as the cut-off frequency f_L. That is, $f_{CS} = f_{L1} = f_L = 1$ kHz.

$$f_{CS} = \frac{1}{2\pi R_{CS} C_S} = \frac{1}{2\pi \times 139 \times C_S} = 1 \text{ kHz or } C_S = 1.15 \text{ mF.}$$

Set the second break frequency corresponding to the next higher resistance. That is, $f_{L2} = f_{C1} = f_L/10 = 1$ kHz / $10 = 100$ Hz

$$f_{L2} = f_{C2} = f_L/10 = 1 \text{ kHz} / 10 = 100 \text{ Hz}$$

$$f_{C2} = \frac{1}{2\pi R_{C2} C_2} = \frac{1}{2\pi \times 12.6 \text{ k}\Omega \times C_2} = 100 \text{ Hz or } C_2 = 0.13 \ \mu F$$

Set the third break frequency corresponding to the highest resistance, $f_{L3} = f_{C1} = f_L/20 = 1$ k/20 = 50 Hz.

$$f_{C1} = \frac{1}{2\pi R_{C1} C_1} = \frac{1}{2\pi \times 1.43 M\Omega \times C_1} = 50 \text{ Hz or } C_1 = 2.23 \text{ }\eta F$$

ADDITIONAL SHUNT CAPACITOR

$V_{GS} = -0.8145$ V.
$V_{DG} = V_{DS} + V_{SG} = V_{DS} - V_{GS} = 3 + 0.8145 = 3.8145$ V.
The zero-biased gate-source capacitance, $C_{gso} = 4$ pF.
The zero-biased gate-drain capacitance, $C_{\mu o} = 4$ pF.
The gate-source junction capacitance can be found from
$C_{gs} = C_{gsoo}/[1 + |V_{GS}|/V_{bi}]^{1/2}$
where V_{bi} the gate junction potential.
Assuming $V_{bi} = 1.0$ V, we get
$C_{gs} = 4pF/[1 + 0.8145/1.0]^{1/2} = 2.97$ pF.
The gate-drain junction capacitance can be found from
$C_{gd} = C_{gdo}/[1 + |V_{GD}|/V_{bi}]^{1/2}$
Assuming $V_{bi} = 1.0$ V, we get
$C_{gd} = 4pF/[1 + 3.8145/1.0]^{1/2} = 1.82$ pF.
$R_{Cgs} = R_s || R_G \equiv 200 \Omega$.
$R_{Cgd} = (R_L || R_D) + (R_s || R_G)[1 + g_m(R_L || R_D)] = 3.22$ kΩ.
The high 3-dB frequency f_H is given by

$$f_H = \frac{1}{2\pi\left[R_{Cgs}C_{gs} + R_{Cgd}(C_{gd} + C_x)\right]} = 250 \text{ kHz or } C_{gd} + C_x = 197.5 \text{ pF}$$

which gives $C_x = 197.5 - 1.82 = 195.7$ pF. This is the additional capacitor C_x that is to be connected between the gate and drain terminals of the JFET.

11.5   EWB SIMULATION

The EWB has circuit models for many types of BJTs and FETs. First, you get the schematic of a BJT from the list of Active devices and an FET from the list of FET devices. Then, you double-click on the device (BJT or JET) and the library of BJTs or FETs will open. The library has models of many commercially available BJTs and FETs. Also, you can change the model parameters by choosing **Edit**.

We will use EWB to simulate BJT and JFET amplifiers to verify the design specifications.

Frequency Response of Common-Emitter Amplifier

The steps to follow are:
1. Open file FIG11_1.CA4 from the EWB file menu.
2. Check that the Function Generator is set to 10 mV, 5 kHz (sine wave). Otherwise, change the settings. Run the simulation. *If the simulation does not reach the steady-state condition, stop the simulation by clicking the stop button.*
3. Check that the Oscilloscope has settings of AC; time base: 0.02 ms/div; channel A: 10 mV/div; channel B: 1 V/div. Make sure that the output is not distorted. Otherwise, reduce the input signal. *If the oscilloscope does not show anything, then change the triggering to 'Auto' instead of 'EXT'.*
4. Set the Bode plotter to magnitude, the gain on LOG (I = 0, F = 50 dB), and the frequency on LOG (I = 10 Hz, F = 10 MHz).
5. Use the cursor on the Bode plotter to measure f_L, f_H, and $A_{v(mid)}$. Record the values in Table 11.1.

Table 11.1

	Calculated	Simulated	Practically Measured
f_L (kHz)	1	1.05 (at 35.7 dB)	
f_H (kHz)	250	631 (at 35.8 dB)	
$A_{v(mid)}$	100	86.1 (or 38.7 dB)	

6. If f_L is not 1 kHz, which capacitor would you adjust to get f_L = 1 kHz? _____. Try C_E and record. C_E = _____ µF for f_L = 1 kHz. You may not get f_L = 1 kHz, but it should be close to this value.
7. If f_H is not 250 kHz, which capacitor would you adjust to get f_H = 250 kHz? _____. Try C_x and record. C_x = _____ µF for f_H = 250 kHz. f_H should be sensitive to any change in the value of C_x.
8. Run the AC parametric sweep by varying C_x = 8.51 pF nominally (connected between the base and collector in Fig. 11.1) from 0 pF to 20 pF with an increment of 10 pF. Set the AC options: start frequency: 100 Hz; end frequency: 10 MHz; sweep type: decade; number of points: 100; and output node: 4. Plot the AC response of the output voltage. Discuss the effects of C_x.

_____
_____
_____

Frequency Response of Common-Source Amplifier

The steps to follow are:
1. Open file FIG11_2.CA4 from the EWB file menu.
2. Check that the Function Generator is set to 100 mV, 5 kHz (sine wave). Otherwise, change the settings. Run the simulation. *If the simulation does not reach the steady-state condition, stop the simulation by clicking the stop button.*
3. Check that the Oscilloscope has settings of AC (if the settings are on DC, change to DC for both channels A and B); time base: 0.02 ms/div; channel A: 100 mV/div; channel B: 500 mV/div. Make sure that the output is not distorted. Otherwise, reduce the input signal.
4. Set the Bode plotter to magnitude, the gain on LOG (I = 0, F = 20 dB), and the frequency on LOG (I = 10 Hz, F = 10 MHz).
5. Use the cursor on the Bode plotter to measure f_L, f_H and $A_{v(mid)}$. Record the values in Table 11.2.
6. If f_L is not 1 kHz, which capacitor would you adjust to get f_L = 1 kHz? _____. Try C_S and record. C_S =_____ (0.45) µF for f_L = 1 kHz.
7. If f_H is not 250 kHz, which capacitor would you adjust to get f_H = 250 kHz? _____. Try C_x and record. C_x = _____ µF for f_H = 250 kHz.

Table 11.2

	Calculated	Simulated	Practically Measured
f_L (kHz)	1	606 Hz (at 12.5 dB)	
f_H (kHz)	250	402 (at 12.2 dB)	
$A_{v(mid)}$	10 (20 dB)	5.82 (or 15.3 dB)	

11.6 WRITE-UP/CONCLUSIONS

Write a brief report, summarizing the results of the design experiment and what you have learned or confirmed about the frequency response of amplifiers.

Comment on the relationship between theory, simulation and practical circuit performance and the precautions that must be taken when using this learning approach.

_____

_____

_____

_____

_____

11.7 REINFORCEMENT EXERCISES

1. A non-inverting amplifier is shown in Fig. 11.3.

Figure 11.3
Non-inverting amplifier

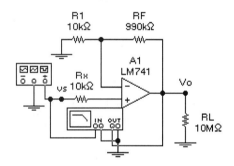

The steps to follow are:

A. Open file FIG11_3.CA4 from the EWB file menu. Run the simulation.

B. Check that the Function Generator is set to 10 mV, 1 kHz. Otherwise, change the settings.

C. Set the Bode plotter to magnitude, the gain on LOG (I = 0, F = 50 dB) and the frequency on LOG (I = 1 Hz, F = 10 MHz).

D. Use the cursor on the Bode plotter to measure f_L, f_H, and $A_{v(mid)}$. Record the values in Table 11.3.

Table 11.3

	Calculated	Simulated	Practically Measured
f_L (kHz)	0	0	
f_H (kHz)	15	14.4	
BW = f_H - f_L (kHz)	15	14.4	
$A_{v(mid)}$	100	100 (or 40 dB)	
GBW = $A_{v(mid)}$ BW (MHz)	1.5	1.44 MHz	

Compare the gain bandwidth product (GBW) with the op-amp GBW of 1.5 MHz. Comment.

_____
_____
_____

E. Run the Bode plot simulation for different values of R_F. Complete Table 11.4.

Table 11.4

R_F (kΩ)	$A_{v(mid)}$		BW (kHz)	GBW (MHz)
990	100		14.4	1440
99	11	10.96 (20.8 dB)	134	13,400
9.9	2	1.99 (5.98 dB)	759	75,900
1	1.1	1.1 (0.83 dB)	1360	136,000

As the gain decrease, the BW decreases. Why? Explain.

_____
_____
_____

2. An inverting integrator is shown is Fig. 11.4.
Steps to follow are:
A. Open file FIG11_4.CA4 from the EWB file menu.

B. Run the Bode plot simulation. *Note that the gain is set to linear here or change it to log.*
C. Record the values in Table 11.5.
D. You will notice that the integrator behaves as a low-pass filter with a low-pass gain of R_F/R_1, and a break frequency of $f_H = 1/(2\pi R_F C_F)$. Why? Explain.

_____

_____

_____

D. (Only for EWB version 5.0) Run the AC parametric sweep by varying $R_F = 6.3$ MΩ nominally (in Fig. 11.4) from 1 MΩ to 50 MΩ with an increment of 20 MΩ. Plot the AC response of the output voltage v_o at node 1. Discuss the effects of R_F.

_____

_____

_____

Figure 11.4
Inverting amplifier

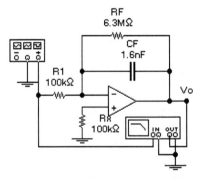

Table 11.5

	Calculated	Simulated	Practically Measured
f_L (Hz)	0	0	
f_H (Hz)	15.8	27	
BW = f_H - f_L (Hz)	15.8	27	
$A_{v(mid)}$	63	63	

3. An inverting differentiator is shown is Fig. 11.5. The steps to follow are:
A. Open file FIG11_5.CA4 from the EWB file menu.
B. Run the Bode plot simulation. Record the values in Table 11.6.

Figure 11.5
Inverting differentiator

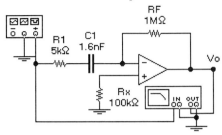

C. You will notice that the differentiator behaves as a high-pass filter with a high-pass gain of $-R_F/R_1$, and a break frequency of $f_L = 1/(2\pi R_1 C_1)$. Why? Explain.

_____

_____

_____

Table 11.6

	Calculated	Simulated	Practically Measured
f_L (kHz)	19.8	19.8 (at 43 dB)	
$A_{v(mid)}$	200	199.5 (or 46 dB)	

11.8 DESIGN PROBLEMS

There may be more than one solution to the following problems. Use EWB or PSPICE to verify your design. Also, build and test, if possible. Determine the voltage and current ratings of active and passive components.

1. Modify the design Prob. 6.2 so that the amplifier operates in the frequency range from 10 Hz to 100 kHz.
2. Modify the design Prob. 6.3 so that the amplifier operates in the frequency range from 5 Hz to 50 kHz.
3. Modify the design Prob. 6.4 so that the amplifier operates in the frequency range from 10 Hz to 50 kHz.
4. Modify the design Prob. 6.5 so that the amplifier operates in the frequency range from 10 Hz to 50 kHz.

12 DESIGN OF MULTI-STAGE AMPLIFIERS FOR FREQUENCY RESPONSE

12.1 LEARNING OBJECTIVES

To design a multi-stage BJT amplifier to give a specified frequency response. We will use Electronics Workbench to verify the design and evaluate the performance of the amplifier.

At the end of this lab, you will
- Be familiar with the frequency characteristic of multi-stage amplifiers.
- Be able to analyze and design multi-stage amplifiers to meet certain frequency specifications.

12.2 THEORY

Two or more amplifiers are often connected in cascade to meet voltage gain, frequency range, input impedance and/or output impedance requirements. A two-stage (CE-CC) amplifier is shown in Fig. 12.1. The CE-stage gives the desired voltage gain and the emitter-follower (CC-stage) gives a low output impedance, while offering a high load impedance to the CE-stage. The short-circuit and the zero-value methods can be applied also to multistage amplifiers. However, the number of time constants will increase with the number of capacitors in the multi-stage amplifier.

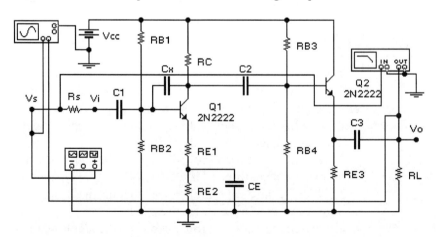

Figure 12.1
Two-stage amplifier

The CC-stage will add a low-break frequency f_{C3} due to C_3, and two high break frequencies due to the internal capacitors of the transistor Q_2.

Considering only the dominant poles of each stage separately, the transfer function of the amplifier can be written approximately in the general form

$$A(s) = \frac{A_{v1} A_{v2} s}{(1 + s/\omega_{L1})(1 + s/\omega_{H1})(1 + s/\omega_{L2})(1 + s/\omega_{H2})} \qquad (12\text{-}1)$$

where A_{v1} and A_{v2} are the mid-frequency gains of the stages. ω_L and ω_H are the dominant low and high cut-off frequencies, respectively.

Alternately, one could find the effective low cut-off frequency from the effective time constant of all capacitors. That is,

$$f_L = \frac{1}{2\pi} \sum_{k=1}^{n} \tau_{Ck} = \frac{1}{2\pi} \sum_{k=1}^{n} \frac{1}{C_k R_{Ck}} \qquad (12\text{-}2)$$

where τ_{Ck} is the time constant due to kth capacitor only, and R_{Ck} is the Thevenin's equivalent resistance presented to C_k. The zero-value method finds the time constant τ_{Cm} for mth capacitor by considering one capacitor at a time while setting other capacitors to 0 (or effectively open circuiting them). Thus, the high 3-dB frequency is determined from the effective time constant of all capacitors for the two transistors. That is,

$$f_H = \frac{1}{2\pi \sum_{m=1}^{} \tau_{Cm}} = \frac{1}{2\pi \sum_{m=1}^{} C_m R_{Cm}} \qquad (12\text{-}3)$$

where τ_{Cm} is the time constant due to mth capacitor only, and R_{Cm} is the Thevenin's equivalent resistance presented to C_m.

Frequency Response of Common-Emitter Stage

The determination of the low- and high cut-off frequencies will be similar to that described in Chapter 11, except the (mid-frequency) input resistance of the second stage will be the load resistance of the first stage. That is,

$$R_{L1} = (R_{B3} \| R_{B4}) \| [r_{\pi 2} + (R_L \| R_{E3})(1 + \beta_F)] \qquad (12\text{-}4)$$

The low cut-off frequency f_{L1} for the first stage can be found from

$$f_{L1} = \frac{1}{2\pi} \left[\frac{1}{C_1 R_{C1}} + \frac{1}{C_2 R_{C2}} + \frac{1}{C_E R_{CE}} \right] \qquad (12\text{-}5)$$

Thus, the high cut-off frequency f_{H1} for the first stage is given by

$$f_{H1} = \frac{1}{2\pi (R_{C\pi 1} C_{\pi 1} + R_{C\mu 1} C_{\mu 1})} \qquad (12\text{-}6)$$

Frequency Response of Emitter-Follower Stage

The configuration of an emitter follower is shown in Fig. 12.2. It has a band-pass frequency characteristic. For the determination of the frequency response, we need to calculate (a) the DC biasing to find the small-signal parameters of the transistors, (b) the mid-frequency voltage gain $A_{v(mid)}$, (c) the low cut-off frequency f_{L2}, and (d) the high cut-off frequency f_{H2}. In Chapter 7, we described how to find the DC biasing and the mid-frequency gain. The output resistance of the first stage will be the source resistance of the second stage. That is, $R_{s1} = R_{o1}$ (R_C.

Figure 12.2
Emitter follower stage

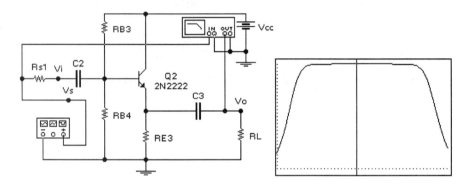

LOW BREAK FREQUENCIES

The low break frequencies depend upon the coupling-capacitors C_2 and C_3. The Thevenin's equivalent resistance presented to C_2 can be found from

$$R_{C2} = R_{s1} + (R_{B3} \| R_{B4}) \| \left[r_{\pi 2} + (1 + \beta_F)(R_{E3} \| R_L) \right] \qquad (12\text{-}7)$$

Thevenin's equivalent resistance presented to C_3 is

$$R_{C3} = R_L + R_o \qquad (12\text{-}8)$$

where R_o is the output resistance given by

$$R_o = R_{E3} \left\| \frac{r_{\pi 2} + (R_{s1} \| R_{B3} \| R_{B4})}{1 + \beta_F} \right. \qquad (12\text{-}9)$$

The low cut-off frequency for the second stage can be found from

$$f_{L2} = \frac{1}{2\pi} \left[\frac{1}{C_2 R_{C2}} + \frac{1}{C_3 R_{C3}} \right] \qquad (12\text{-}10)$$

HIGH BREAK FREQUENCIES

The resistance faced by $C_{\pi 2}$ is given by,

$$R_{C\pi 2} = \left(R_{s1}\|R_{B3}\|R_{B4}\right)\|\left[r_{\pi 2} + \left(1+\beta_F\right)\left(R_{E3}\|R_L\right)\right] \qquad (12\text{-}11)$$

Thevenin's equivalent resistance faced by $C_{\mu 2}$ is given by

$$R_{C\mu 2} = r_{\pi 2} \left\| \frac{\left(R_{s1}\|R_{B3}\|R_{B4}\right) + \left(R_{E3}\|R_L\right)}{1 + g_{m2}\left(R_L\|R_{E3}\right)} \right. \qquad (12\text{-}12)$$

The high cut-off frequency f_{H2} is given by

$$f_{H2} = \frac{1}{2\pi\left(R_{C\pi 2}C_{\pi 2} + R_{C\mu 2}C_{\mu 2}\right)} \qquad (12\text{-}13)$$

12.3 READING ASSIGNMENT

Study chapter(s) on multi-stage amplifiers.

12.4 ASSIGNMENT

Design Specifications

We will design a multi-stage amplifier shown in Fig. 12.1. The design specifications are:
- Midband voltage gain, $|A_{mid}| = |v_o/v_s| = 100$.
- Low cut-off frequency, $f_L = 1$ kHz.
- High cut-off frequency, $f_H = 250$ kHz.
- Load resistance, $R_L = 10$ kΩ.
- Source resistance, $R_s = 200$ Ω.
- Use NPN-transistor 2N2222 whose $\beta_F = 200$.
- Quiescent collector current, $I_{C1} = 5$ mA (assumed).
- $r_{\pi 1} = \beta_F V_T/I_{C1} = 200 \times 25.8$ mV/5 mA $= 1032$ Ω.
- $g_{m1} = \beta_F/r_{\pi 1} = 200/1032 = 193.8$ mA/V.
- Quiescent collector current, $I_{C2} = 2$ mA (assumed).
- $r_{\pi 2} = \beta_F V_T/I_{C2} = 200 \times 25.8$ mV/2 mA $= 2580$ Ω.
- $g_{m2} = \beta_F/r_{\pi 2} = 200/2580 = 77.52$ mA/V.

COMMON-EMITTER STAGE

Design the common-emitter stage to meet the following specifications:
- Midband voltage gain, $|A_v| = v_{o1}/v_s = 100$.
- Low cut-off frequency, $f_{L1} = 1$ kHz.
- High cut-off frequency, $f_{H1} = 250$ kHz.

From the values in Chapters 5 and 11 of this lab manual, $R_C = 1.6$ kΩ, $R_{B1} = 68$ kΩ, $R_{B2} = 21$ kΩ, $R_E = 796$ Ω, $R_{E1} = 9$ Ω, and $R_{E2} = 787$ Ω. Capacitances are given by $C_E = 10.73$ µF, $C_1 = 0.61$ µF, $C_2 = 0.274$ µF, and $C_x = 8.51$ pF (from Chapter 11).

$$R_{B1} \| R_{B2} = 68 \text{ k} \| 21 \text{ k} = 16 \text{ k}\Omega$$

$$R_{L1} = (R_{B3} \| R_{B4}) \| [r_{\pi 2} + (R_L \| R_{E3})(1 + \beta_F)] = 87.4 \text{ k}\Omega$$

$$R_{C2} = R_C + R_{L1} = 89 \text{ k}\Omega$$

$$f_{C2} = \frac{1}{2\pi R_{C2} C_2} = \frac{1}{2\pi \times 89 \text{ k}\Omega \times C_2} = 6.5 \text{ Hz}$$

which is low enough and will not affect the design for the low cut-off frequency of 1 kHz.

EMITTER FOLLOWER STAGE

Design the emitter follower stage shown in Fig. 12.2 to meet the following specifications:
- Midband voltage gain, $|A_{v2}| = v_o/v_{o1} = 1$.
- Low cut-off frequency, $f_{L2} < 1$ kHz.
- High cut-off frequency, $f_{H2} > 250$ kHz.
- Load resistance, $R_L = 10$ kΩ.

From the values in Chapters 7 of this lab manual, $R_{E3} = 5$ kΩ, $R_{B3} = 188$ kΩ, $R_{B4} = 216$ kΩ and $R_{s1} = R_C = 1.6$ kΩ.

Assuming $C\pi \cong C_{jeo} = 19.5$ pF and $C_\mu \cong C_{\mu o} = 9.63$ pF for transistor Q_2, the unity-gain frequency of the transistor Q_2 can be found approximately as

$$f_T = \frac{g_{m2}}{2\pi(C_\pi + C_\mu)} = \frac{77.52 \text{ mA/V}}{2\pi(19.5 \text{ pF} + 9.63 \text{ pF})} = 423.5 \text{ MHz}$$

Since this is higher than $f_{H2} > 250$ kHz, there is no need for setting the upper frequency.

$$R_o = R_{E3} \left\| \frac{r_{\pi 2} + (R_{s1} \| R_{B3} \| R_{B4})}{1 + \beta_F} \right\| = 20.7 \text{ } \Omega$$

$$R_{C3} = R_L + R_o = 10.02 \text{ k}\Omega$$

Let us set the fourth break frequency due to C_3 at $f_{L2} = f_{C3} = f_L/100 = 1\text{k}/20 = 10$ Hz.

$$f_{C2} = \frac{1}{2\pi R_{C31} C_3} = \frac{1}{2\pi \times 10.02 \text{ k}\Omega \times C_3} = 10 \text{ Hz or } C_2 = 1.56 \text{ } \mu F$$

12.5 EWB SIMULATION

Frequency Response of Emitter Follower

We will use EWB to simulate the multi-stage amplifier to verify the design specifications.

The steps to follow are:
1. Open file FIG12_2.CA4 from the EWB file menu.
2. Check that the Function Generator is set to 1 V, 5 kHz (sine wave). Otherwise, change the settings. Run the simulation.
3. Check that the Oscilloscope has settings of AC; time base: 0.02 ms/div; channel A: 1 V/div; channel B: 1 V/div. Make sure that the output is not distorted. Otherwise, reduce the input signal. *If the oscilloscope does not show anything, then change the triggering to 'Auto' instead of 'EXT'.*
4. Set the Bode plotter to magnitude, the gain on LIN (I = 0, F = 1), and the frequency on LOG (I = 1 Hz, F = 100 MHz).
5. Use the cursor on the Bode plotter to measure f_{L2}, f_{H2} and $A_{v2(mid)}$. Record the values in Table 12.1.

Table 12.1

	Calculated	Simulated	Practically Measured
f_{L2} (Hz)	10	~8.2 Hz	
f_{H2} (MHz)	423.5	~39.8	
BW = f_{H2} - f_{L2} (MHz)	423.5	~39.8 MHz	
$A_{v2(mid)}$	1	0.98	

Frequency Response of Multi-Stage Amplifier

The steps to follow are:
1. Open file FIG12_1.CA4 from the EWB file menu.
2. Check that the Function Generator is set to 10 mV, 5 kHz (sine wave). Otherwise, change the settings. Run the simulation. *If the simulation does not reach the steady-state condition, stop the simulation by clicking the stop button.*
3. Check that the Oscilloscope has settings of AC; time base: 0.02 ms/div; channel A: 10 mV/div; channel B: 1 V/div. Make sure that the output is not distorted. Otherwise, reduce the input signal. *If the oscilloscope does not show anything, then change the triggering to 'Auto' instead of 'EXT'.*
4. Set the Bode plotter to magnitude, the gain on LIN (I = 0, F = 100), and the frequency on LOG (I = 100 Hz, F = 10 MHz).
5. Use the cursor in the Bode plotter to measure f_L, f_H and $A_{v(mid)}$. Record the values in Table 12.2.

6. If f_L is not 1 kHz, which capacitor would you adjust to get f_L = 1 kHz? _____. Try C_E and record. C_E = _____ (6) µF for f_L = 1 kHz.

7. If f_H is not 250 kHz, which capacitor would you adjust to get f_H = 250 kHz? _____. Try C_x and record. C_x = _____ (50) pF for f_H = 250 kHz.

8. (Only for EWB version) Run the AC parametric sweep by varying C_x = 8.51 pF nominally (in Fig. 12.1) from 1 pF to 50 pF with an increment of 10 pF. Set the AC options: start frequency: 100 Hz; end frequency: 10 MHz; sweep type: decade; number of points: 100; and output node: 2. Plot the AC response of the output voltage. Discuss the effects of C_x.

_____

_____

_____

Table 12.2

	Calculated	Simulated	Practically Measured
f_L (kHz)	1	~0.585 1.07 kHz (at 68)	
f_H (kHz)	250	~1080 575 kHz (at 68)	
BW = f_H - f_L (kHz)	249	1079.4 574 kHz	
$A_{v(mid)}$	100	98	

12.6 WRITE-UP/CONCLUSIONS

Write a brief report, summarizing the results of the design experiment and what you have learned or confirmed about the frequency response of multi-stage amplifiers.

Comment on the relationship between theory, simulation and practical circuit performance and the precautions that must be taken when using this learning approach.

_____

_____

_____

12.7 REINFORCEMENT EXERCISES

A multi-stage amplifier consisting of two identical common-emitter stages is shown in Fig. 12.3.

The steps to follow are:
1. Open file FIG12_3.CA4 from the EWB file menu.
2. Check that the Function Generator is set to 100 µV, 5 kHz. Otherwise, change the settings. Run the simulation.
3. Set the Bode plotter to magnitude, the gain on LOG (I = 0, F = 80 dB), and the frequency on LOG (I = 10 Hz, F = 10 MHz).
4. Use the cursor in the Bode plotter to measure f_L, f_H, and $A_{v9mid)}$. Record the values in Table 12.3.

Table 12.3

	Calculated	Simulated	Practically Measured
f_L (kHz)	1	~1.38 (at 70.8 dB)	
f_H (kHz)	485	~402 (at 70.8 dB)	
BW = f_H - f_L (kHz)	484	~~401 kHz	
$A_{v(mid)}$*	10^4 (80 dB)	4955 (73.9 dB)	

* Note: From the oscilloscope, we get 0.8731V/0.0002 = 4365 (72.8 dB).

5. Connect a capacitor C_x = 8.51 pF between the base and collector of Q_1. Run the simulation. Record: f_H ___316kHz___, which is lower than the value without C_x. Why? Explain.

6. Disconnect C_x.
7. Connect the input of the Bode plotter to the input signal v_s and the output of the Bode plotter to the output of the first stage. Set the Bode

plotter to magnitude, the gain on LOG (I = 0, F = 50 dB), and the frequency on LOG (I = 10 Hz, F = 10 MHz). Run the simulation.

Figure 12.3
Two stage common-emitter amplifier

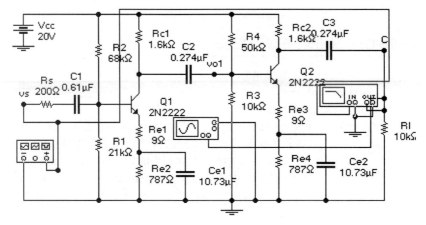

8. Use the cursor on the Bode plotter to measure f_{L1}, f_{H1} and $A_{v1(mid)}$. Record the values in Table 12.4.

Table 12.4

	Calculated	Simulated	Practically Measured
f_{L1} (kHz)	1	0.913	
f_{H1} (kHz)	485	~590	
$A_{v1(mid)}$	100 (40 dB)	62.4 (35.9 dB)	

9. Connect the input of the Bode plotter to the output of the first stage and the output of the Bode plotter to the output of the second stage. Set the Bode plotter to magnitude, gain on LOG (I = 0, F = 50 dB), and frequency on LOG (I = 10 Hz, F = 100 MHz). Run the simulation.
10. Use the cursor on the Bode plotter to measure f_{L2}, f_{H2}, and $A_{v2(mid)}$. Record the values in Table 12.5.

Table 12.5

	Calculated	Simulated	Practically Measured
f_{L2} (1Hz)	1	0.872	
f_{H2} (kHz)	485	27.5 MHz	
$A_{v2(mid)}$	100 (40 dB)	79.4 (38 dB)	

Compare Tables 12.4 and 12.5. Although, the two stages are identical, their gains differ. Why? Explain.

_____

_____

_____

11. Derive the transfer function of the amplifier in the form as follows from Tables 12.4 and 12.5.

$$A(s) = \frac{(A_{v1}) \times (A_{v2}) s}{\left[1 + s/(\omega_{L1})\right]\left[1 + s/(\omega_{H1})\right]\left[1 + s/(\omega_{L2})\right]\left[1 + s/(\omega_{H2})\right]}$$

From Table 12.3,

$$A(s) = \frac{(A_{v(mid)}) s}{\left[1 + s/(\omega_L)\right]\left[1 + s/(\omega_H)\right]}$$

12.8 DESIGN PROBLEMS

There may be more than one solution to the following problems. Use EWB or PSPICE to verify your design. Also, build and test, if possible.

1. Modify the design Prob. 7.4 so that the amplifier operates in the frequency range from 10 kHz to 80 kHz.
2. Modify the design Prob. 7.5 so that the amplifier operates in the frequency range from 20 kHz to 60 kHz.

13 DESIGN OF ACTIVE FILTERS

13.1 LEARNING OBJECTIVES

To design active filters to give a specified frequency response. We will use Electronics Workbench to verify the design and evaluate the performance of active filters.

After the end of this lab, you will
- Be familiar with the types and frequency characteristics of active filters.
- Be able to analyze and design active filters.

13.2 THEORY

A filter is a frequency selective circuit that passes a specified band of frequencies, and blocks, or attenuates signals of frequencies outside this band. An active filter consists of an op-amp (s), resistors and capacitors. The op-amp is used for voltage gain, and also offers a high input impedance and a low output impedance. Active filters can be classified as low-pass, high-pass, band-pass, and band-reject. A low pass filter is shown in Fig. 13.1. The transfer function is

$$H(j\omega) = \frac{V_o(j\omega)}{V_s(j\omega)} = \frac{K_L}{1 + j\omega RC} \qquad (13\text{-}1)$$

which gives the high cut-off (3-dB) frequency f_H as

$$f_H = \frac{1}{2\pi RC} \qquad (13\text{-}2)$$

Figure 13.1
Low-pass filter

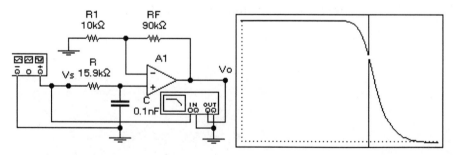

The low frequency gain K_L is given by

$$K_L = \left(1 + \frac{R_F}{R_1}\right) \qquad (13\text{-}3)$$

148

A high-pass filter can be formed by interchanging the frequency dependent-resistor and capacitor of the low-pass filter. This is shown in Fig. 13.2. The transfer function is

$$H(j\omega) = \frac{V_o(j\omega)}{V_s(j\omega)} = \frac{j\omega K_H}{j\omega + 1/RC} \quad (13\text{-}4)$$

which gives the low cut-off (or 3-dB) frequency f_H as

$$f_L = \frac{1}{2\pi RC} \quad (13\text{-}5)$$

where the high-frequency gain K_H is given by

$$K_H = \left(1 + \frac{R_F}{R_1}\right) \quad (13\text{-}6)$$

Figure 13.2
High-pass filter

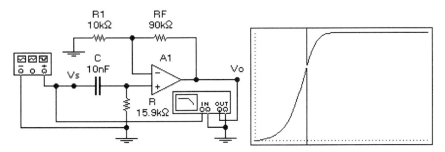

Figure 13.3
Band-pass filter

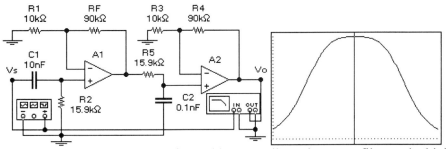

A band-pass filter can be formed by cascading a low-pass filter and a high-pass filter. This is shown Fig. 13.3. The transfer function is given by

$$H(j\omega) = \frac{V_o(j\omega)}{V_s(j\omega)} = \frac{j\omega K_H K_L}{(1 + j\omega R_2 C_1)(j\omega + 1/R_5 C_2)} \quad (13\text{-}7)$$

The roll-off of a first-order filter is only -20dB/decade in the stop-band. A second-order filter exhibits a stop-band fall characteristic of -40 dB/decade roll-off, and it is preferable to the first-order filter. In addition, a second-order filter can be the building block for higher-order filters (i.e., n = 4, 6, ..) and is commonly used as such.

Second-Order Low-Pass Filter A second-order filter, which can give both Butterworth response (that is, a flat response) and a voltage gain, is shown in Fig. 13.4. This circuit is also known as the *Sallen-Key* circuit whose transfer function is given by

$$H(s) = \frac{V_o(s)}{V_s(s)} = \frac{\left(K_L / R_2 R_3 C_1 C_2\right)}{\left(s^2 + s \dfrac{R_3 C_2 + R_2 C_2 + R_1 C_2 - K_L R_2 C_1}{R_2 R_3 C_1 C_2} + \dfrac{1}{R_2 R_3 C_1 C_2}\right)} \quad (13\text{-}8)$$

where $K_L = (1 + R_F/R_1)$ is the low-frequency gain.

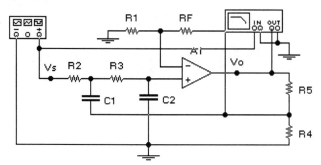

Figure 13.4 Second-order low-pass filter

Normally, $R_1 = R_2 = R_3 = R$, $C_1 = C_2 = C$. Then, the cut-off frequency f_H becomes

$$f_H = \frac{\omega_o}{2\pi} = \frac{1}{2\pi\sqrt{R_2 R_3 C_1 C_2}} = \frac{1}{2\pi RC} \quad (13\text{-}9)$$

The transfer function is reduced to

$$H(s) = \frac{K_L \omega_o^2}{s^2 + (3 - xK_L)\omega_o s + \omega_o^2} \quad (13\text{-}10)$$

For a flat response,

$$xK_L = x\left(1 + \frac{R_F}{R_1}\right) = 3 - \sqrt{2} = 1.586 \quad (13\text{-}11)$$

where x is the fraction of the output voltage that is fed back through the capacitor C_1, and it is given by

$$x = \frac{R_4}{R_4 + R_5} \quad (13\text{-}12)$$

The quality factor Q is related to K_L by

$$Q = \frac{1}{3 - xK_L} \quad (13\text{-}13)$$

Thus, for Q = 0.707, x K_L = 1.586. That is, K_L = 1.586/x, which allows more gain K_L by choosing a lower value of x (<1).

Second-Order High-Pass Filter   A second-order high-pass filter can be formed from a second-order low-pass filter by interchanging the frequency dominant resistors and capacitors. This is shown in Fig. 13.5.

Figure 13.5
Second-order high-pass filter

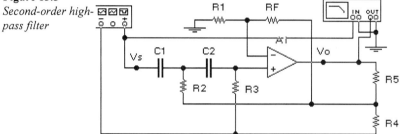

13.3   READING ASSIGNMENT

Study chapter(s) on active filters.

13.4   ASSIGNMENT

Design Specifications   We will design a second-order low-pass filter and a high-pass filter, and then cascade them to get the band-pass characteristics. The design specifications are:
- Pass-band voltage gain, $K_{BP} = K_H K_L = v_o/v_s = 100$.
- Low cut-off (or 3-dB) frequency, f_L = 1 kHz.
- High cut-off (or) 3-dB frequency, f_H = 100 kHz.
- Use ideal op-amps.
- DC supply voltages, $V_{CC} = -V_{EE}$ = 20 V.

Low-Pass Filter   Design the second-order low-pass filter shown in Fig. 13.4 to meet the following specifications:
- Low-pass voltage gain, $K_L = v_{o1}/v_s$ = 10.
- High cut-off frequency, f_H = 100 kHz.

RESISTORS AND CAPACITORS

Let $R_2 = R_3 = R$, and $C_1 = C_2 = C$. Choose C = 0.1 nF. The value of R becomes

$$R = \frac{1}{2\pi f_H C} = \frac{1}{2\pi \times 100 \text{ kHz} \times 0.1 \text{ nF}} = 15.9 \text{ k}\Omega$$

Let R_1 = 15.9 kΩ. For flat response, $1 + R_F/R_1$ = 1.586 which gives R_F = (1.586 - 1) R_1 = 9.3 kΩ.

For K_L = 10, x = 1.586/K_L = 1.586/10 = 0.1586. Thus,

151

$$\frac{R_5}{R_4} = \frac{1}{x} - 1 = \frac{1-x}{x} \qquad (13\text{-}14)$$

which, for $x = 0.1586$ and $R_4 = 2$ kΩ (say), gives $R_5 = 10.5$ kΩ.

High-Pass Filter Design the second-order high-pass filter shown in Fig. 13.5 to meet the following specifications:
- High-pass voltage gain, $K_H = v_o/v_{o1} = 10$.
- Low cut-off (or 3-dB) frequency, $f_L = 1$ kHz.

RESISTORS AND CAPACITORS

Let $R_2 = R_3 = R$, and $C_1 = C_2 = C$. Choose $C = 10$ nF. The value of R becomes

$$R = \frac{1}{2\pi f_L C} = \frac{1}{2\pi \times 1 \text{ kHz} \times 10 \text{ nF}} = 15.9 \text{ k}\Omega$$

Choosing $R_1 = 15.9$ kΩ, $R_F = (1.586 - 1) R_1 = 9.3$ kΩ. For $K_H = 10$, $x = 1.586/K_H = 1.586/10 = 0.1586$. Choosing $R_4 = 2$ kΩ, then $R_5 = 10.5$ kΩ.

13.5 EWB SIMULATION

We will not simulate the first-order filters which are to used only to illustrate the filter characteristics. The second-order filter exhibit faster rise and fall characteristics than the first-order filters. We will use EWB to simulate second-order active filters using ideal op-amps to verify the design specifications.

Frequency Response of Low-Pass Filter

The steps to follow are:
1. Open file FIG13_4.CA4 from the EWB file menu.
2. Check that the Function Generator is set to 1 V, 10 kHz (sine wave). Otherwise, change the settings.
3. Set the Bode plotter to magnitude, the gain on LIN (I = 0, F = 50), and the frequency on LOG (I = 10 Hz, F = 1 MHz). Run the simulation.
4. Use the cursor on the Bode plotter to measure f_H and K_L. Record the values in Table 13.1.

Table 13.1

	Calculated	Simulated	Practically Measured
f_H (kHz)	100	100 (at 7.8 gain)	
BW = f_H (kHz)	100	100	
K_L	10	11	

Frequency Response of High-Pass Filter

The steps to follow are:
1. Open file FIG13_5.CA4 from the EWB file menu.
2. Check that the Function Generator is set to 1 V, 10 kHz (sine wave). Otherwise, change the settings.

Table 13.2

	Calculated	Simulated	Practically Measured
f_L (kHz)	1	1 (at 7.7 gain)	
K_H	10	11	

3. Set the Bode plotter to magnitude, the gain on LIN (I = 0, F = 50), and the frequency on LOG (I = 10 Hz, F = 1 MHz). Run the simulation.
4. Use the cursor on the Bode plotter to measure f_L and K_H. Record the values in Table 13.2.

Frequency Response of Band-Pass Filter

A band-pass filter is formed by cascading the low-pass filter and the high-pass filter. This is shown in Fig. 13.6.

The steps to follow are:
1. Open file FIG13_6.CA4 from the EWB file menu.
2. Check that the Function Generator is set to 1 V, 10 kHz (sine wave). Otherwise, change the settings.
3. Set the Bode plotter to magnitude, the gain on LOG (I = 0, F = 50), and the frequency on LOG (I = 100 Hz, F = 1 MHz). Run the simulation.
4. Use the cursor on the Bode plotter to measure f_L, f_H, and K_{BP}. Record the values in Table 13.3.

Table 13.3

	Calculated	Simulated	Practically Measured
f_L (kHz)	1	1 (at 38.4 dB)	
f_H (kHz)	100	107 (at 38.4 dB)	
BW = f_H - f_L (kHz)	99	106	
K_{BP}	100 (40 dB)	117.5 (41.4 dB)	

Figure 13.6
Second-order band-pass filter

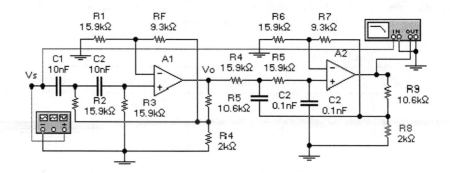

5. (Only for EWB version 5.0) Run the worst-case AC analysis for the minimum and maximum output voltages of the filter in Fig. 13.6 at node 1 (the output of op-amp A_2). Set tolerance: 20%; output node: 1; and Min/Max voltage. Set the AC options: start frequency: 100 Hz; end frequency: 10 MHz; sweep type: decade; number of points: 100. Plot the minimum and maximum output voltages and discuss the effects of tolerance.

_____
_____
_____

6. (Only for EWB version 5.0) Run the AC parametric sweep by varying $R_F = 27.5$ kΩ nominally (in Fig. 13.6) from 10 kΩ to 100 kΩ with an increment of 50 kΩ. Plot the AC response of the output voltage v_o at node 1. Discuss the effects of R_F.

_____
_____
_____

13.6 WRITE-UP/CONCLUSIONS

Write a brief report, summarizing the results of the design experiment and what you have learned or confirmed about the frequency response of active filters.

Comment on the relationship between theory, simulation and practical circuit performance and the precautions that must be taken when using this learning approach.

_____

_____

_____

_____

13.7 REINFORCEMENT EXERCISES

We will compare the characteristics of the low-pass filter in Fig. 13.1 with an ideal op-amp and a practical one.

The steps to follow are:
1. Open file FIG13_1.CA4 from the EWB file menu.
2. Check that the Function Generator is set to 1 V, 10 kHz (sine wave). Otherwise, change the settings.
3. Set the Bode plotter to magnitude, the gain on LIN (I = 0, F = 10), and the frequency on LOG (I = 1 Hz, F = 50 MHz). Run the simulation.
4. Use the cursor on the Bode plotter to measure f_H and K_L. Record the values in Table 13.4 for an ideal op-amp.

The op-amp (ideal one) in the low-pass filter in Fig. 13.1 is replaced by the LM741 op-amp whose particulars are: open-loop gain $A_o = 2 \times 10^2$, the input resistance $R_i = 2$ MΩ, slew rate = 0.5 V/µs, and the unity gain bandwidth $f_u = 1.5$ MHz. This is shown in Fig. 13.7.

The steps to follow are:
1. Open file FIG13_7.CA4 from the EWB file menu.
2. Check that the Function Generator is set to 1 V, 10 kHz (sine wave). Otherwise, change the settings.
3. Set the Bode plotter to magnitude, the gain on LIN (I = 0, F = 50), and the frequency on LOG (I = 1 Hz, F = 50 MHz). Run the simulation.
4. Use the cursor on the Bode plotter to measure f_H and K_L. Record the values in Table 13.4 for the practical op-amp.

Figure 13.7
Low-pass filter with practical op-amp

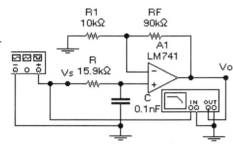

Table 13.4

	Calculated	Simulated for an ideal op-amp*	Simulated for a practical op-amp
f_H (kHz)	100	84.6 (at 7.6 gain)	63.4 (at 7.8 gain)
K_L	10	10	10

Note: In an ideal op-amp model, EWB does not use an op-amp having an finite voltage gain, an infinite input resistance and a zero output resistance. As a result, f_H is not 100 kHz as expected under ideal conditions.

Compare the results in Table 13.4. The f_H has been reduced by the insertion of a practical op-amp. Why? Explain.

_____

_____

_____

A narrow-band filter is shown in Fig. 13.8. The center frequency ω_C, the Q-factor, and the band-pass gain K_{BP} can be found from

$$\omega_C = \frac{1}{2\pi\sqrt{(R_1 \| R_b)C_1 C_2}} \qquad (13\text{-}15)$$

$$Q = \frac{1}{2}\sqrt{\frac{R_2}{R_1}} \qquad (13\text{-}16)$$

$$K_{BP} = \frac{R_2}{2R_1} = 2Q^2 \qquad (13\text{-}17)$$

Figure 13.8
Narrow-band filter

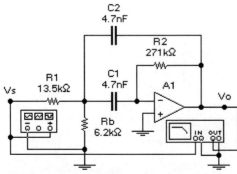

The steps to follow are:
1. Open file FIG13_7.CA4 from the EWB file menu.
2. Check that the Function Generator is set to 1 V, 1 kHz (sine wave). Otherwise, change the settings.

3. Set the Bode plotter to magnitude, the gain on LIN (I = 0, F = 10), and the frequency on LOG (I = 100 Hz, F = 10 kHz). Run the simulation.
4. Use the cursor on the Bode plotter to measure f_H and K_{BP}. Record the values in Table 13.5.

Table 13.5

	Calculated	Simulated	Practically Measured
f_C (Hz)	998	1000	
Q= BW = f_H - f_L (kHz)	2.24	1.26 - 0.794 = 0.466 kHz	
K_{BP}	10	10	

13.8 DESIGN PROBLEMS

There may be more than one solution to the following problems. Use EWB or PSPICE to verify your design. Also, build and test, if possible. Determine the voltage and current ratings of active and passive components.

1. Design a second-order Butterworth band-pass filter to give cut-off frequencies f_L = 400 Hz, f_H = 10 kHz, and a band-pass gain of K_{BP} = 20.
2. Design a third-order Butterworth band-pass filter to give cut-off frequencies f_L = 400 Hz, f_H = 10 kHz, and a band-pass gain of K_{BP} = 20.
3. Design a fourth-order Butterworth band-pass filter to give cut-off frequencies f_L = 400 Hz, f_H = 10 kHz, and a band-pass gain of K_{BP} = 20.

14 DESIGN OF FEEDBACK AMPLIFIERS

14.1 LEARNING OBJECTIVES

To design feedback amplifiers to give a specified bandwidth, gain, input or output impedance. We will use Electronics Workbench to verify the design and evaluate the performance of feedback amplifiers.

At the end of this lab, you will:
- Be familiar with the types and characteristics of feedback amplifiers.
- Be able to analyze and design feedback amplifiers.

14.2 THEORY

You might have realized that the op-amp circuits in Chapter 10 use negative feedback. The amplifier gain A_f is almost independent of the op-amp gain A, but it depends upon the external circuit elements only. For example, the voltage gain of the non-inverting amplifier in Fig. 10.1 is $(1+ R_F/R_1)$, which is independent of the op-amp gain A, and its input impedance is very large. The gain of an inverting amplifier in Fig. 10.2 is $-R_F/R_1$, and the input impedance is approximately R_1. The output impedance of both amplifiers is very small.

In negative feedback, the feedback signal is of opposite polarity, or out of phase by 180° with respect to the input signal. The major benefits of negative feedback in an amplifier are: (1) it stabilizes the overall gain of the amplifier with respect to parameter variations due to temperature, supply voltage, etc., (2) it increases or decreases the input and output impedances, (3) it reduces the distortion and the effect of nonlinearity, and (4) it increases the bandwidth. However, the overall gain is reduced in almost direct proportion to the benefits, and it is often necessary to compensate the decrease in gain by adding an extra amplifier stage. The closed-loop gain A_f is given by

$$A_f = \frac{A}{1+\beta A} \tag{14-1}$$

where A is the gain without any feedback, and β is the feedback factor. For $\beta A \gg 1$, which is usually the case, A_f can be found approximately from

$$A_f = \frac{1}{\beta} \tag{14-2}$$

The bandwidth with feedback, BW_f can be found from

$$BW_F = BW(1+\beta A) \tag{14-3}$$

where BW is the bandwidth without feedback. The gain-bandwidth product remains constant. That is,

$$A_f \times BW_f = A \times BW \qquad (14\text{-}4)$$

Either the voltage or the current can be the input or the output signal. If the output voltage is the feedback signal, then it can be compared with either the input voltage to generate the error voltage signal or the input current to generate the error current signal. Similarly, the output current can be fed back and compared with either the input voltage to generate the error voltage signal or the input current to generate the error current signal. Therefore, there are four configurations depending upon whether the output and feedback signals are voltages or currents.

- Shunt-Shunt (*voltage-sensing current-comparing*) Feedback
- Shunt-Series (*current-sensing current-comparing*) Feedback
- Series-Shunt (*voltage-sensing voltage-comparing*) Feedback
- Series-Series (*current-sensing voltage-comparing*) Feedback

Table 14.1 Feedback Relationships

	Gain A_f	Input Resistance R_{if}	Output Resistance R_{of}
Without Feedback	A	R_i	R_o
shunt-shunt A (V/A) Ω β (A/V) mho	$\dfrac{A}{1+\beta A}$	$\dfrac{R_i}{1+\beta A}$	$\dfrac{R_o}{1+\beta A}$
series-shunt A (V/V) β (V/V)	$\dfrac{A}{1+\beta A}$	$R_i(1+\beta A)$	$\dfrac{R_o}{1+\beta A}$
series-series A (A/V) mho β (V/A) Ω	$\dfrac{A}{1+\beta A}$	$R_i(1+\beta A)$	$R_o(1+\beta A)$
shunt-series A (A/A) β (A/A)	$\dfrac{A}{1+\beta A}$	$\dfrac{R_i}{1+\beta A}$	$R_o(1+\beta A)$

There are two circuits in a feedback amplifier: the amplifier circuit (or A-circuit), and the feedback circuit (or β-circuit). The effective gain is always decreased by a factor of $(1 + \beta A)$. In series type, both A and β circuits are connected in series, and the effective resistance is increased by a factor of $(1 + \beta A)$. In shunt type, A and β circuits are connected in parallel, and the effective resistance is decreased by a factor of $(1 + \beta A)$. The effects of different types of feedback are summarized in Table 14.1. Depending on the type of feedback, the amplifier is normally represented by one of the four amplifier topologies: voltage, current, transconductance, or transresistance. A in Eq. (14-1) is simply a gain, and it could be a voltage gain, current gain, transconductance, or transresistance of the amplifier under the open-loop condition. Thus, A's units could be V/V, A/A, A/V, or V/A.

Series-Shunt Feedback

A series-shunt feedback increases the input resistance and reduces the output resistance. Also, feedback can virtually eliminate distortion nonlinearity. Let us consider the class-B amplifier in Fig. 81, which has a dead-zone in the transfer characteristic, and apply series-shunt feedback. This is shown in Fig. 14.1(a). The op-amp is connected in the non-inverting mode with series-shunt feedback. The output voltage is converted to a feedback voltage V_f by the resistors R_F and R_1 so that $\beta = R_1/(R_1 + R_F)$. V_f is then compared with the input voltage V_s to give the error voltage $V_e = V_s - V_f$. Thus, the closed-loop gain A_f becomes

$$A_f = \frac{V_o}{V_s} = \frac{1}{\beta} = 1 + \frac{R_F}{R_1} \text{(V/V)} \tag{14-5}$$

Therefore, we get the output voltage V_o as

$$V_o = \left(1 + \frac{R_F}{R_1}\right) V_s \tag{14-6}$$

which is the same relation that we obtained for the noninverting amplifier.

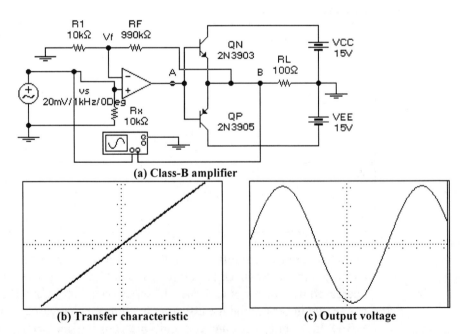

Figure 14.1
Class B amplifier with series-shunt feedback

(a) Class-B amplifier

(b) Transfer characteristic

(c) Output voltage

With this arrangement, either Q_P or Q_N will be on if v_s and v_o differ by $\pm V_{BE}/A$, where A is the open-loop gain of the op-amp. Thus, for $A = 10^5$ and $V_{BE} = 0.7$, the dead-zone should be reduced to less than $\pm 0.7/10^5 = \pm 7$ μV.

The transfer characteristic is shown in Fig. 14.1(b), and the output voltage shown in Fig. 14.1(c) is almost distortion free. $R_x = (R_1 || R_F)$.

Let us make a two-stage common-emitter amplifier by cascading two CE-amplifiers of the configuration shown Fig. 11.1 and then apply series-shunt feedback. This is shown in Fig. 14.2, and it should increase the input resistance, reduce the output resistance, and widen the bandwidth. The capacitor C_F and resistor R_F form the feedback network. The value of C_F should be such that it is virtually shorted over the frequency range of the amplifier. Feedback is taken from the second stage because the output is in phase with the input voltage v_s. Since the gain of the amplifier is $A = 10^4$, the closed loop gain will be almost independent of A, and it can be found from

$$A_f = \frac{V_o}{V_s} = \frac{1}{\beta} = 1 + \frac{R_F}{R_1} \text{ (V / V)} \qquad (14\text{-}7)$$

Figure 14.2
Two-stage common-emitter amplifier with series-shunt feedback

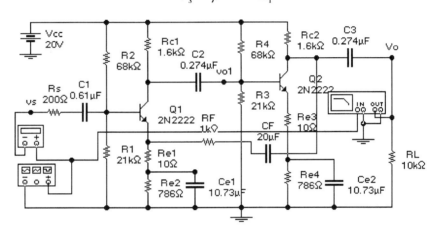

Shunt-Shunt Feedback

A shunt-shunt feedback reduces both the input resistance and the output resistance. Also, this type of feedback can virtually eliminate distortion non-linearity. Let us apply shunt-shunt feedback to the class-B amplifier in Fig. 8.1. This is shown in Fig. 14.3(a). The op-amp is connected in the inverting mode with shunt-shunt feedback. The output voltage is converted to a feedback current I_f by the resistor R_F so that $\beta = -1/R_F$. I_f is then compared with the input current $I_s (= V_s/R_1)$ to give the error current, $I_e = I_s - I_f$. Thus the closed-loop gain A_f becomes

$$A_f = \frac{V_o}{I_s} = \frac{1}{\beta} = -R_F \text{ (V / A)} \qquad (14\text{-}8)$$

Therefore, we get the output voltage V_o

$$V_o = -R_F I_S = \frac{-R_F}{R_1} V_s \qquad (14\text{-}9)$$

161

which is the same relation that we obtained for the inverting amplifier. Either Q_P or Q_N will be on if v_s and v_o differ by $\pm V_{BE}/A$, where A is the open-loop gain of the op-amp. Thus, for $A = 10^5$ and $V_{BE} = 0.7$, the dead-zone will be reduced to less than $\pm 0.7/10^5 = \pm 7$ μV. The transfer characteristic is shown in Fig. 14.3(b), and the output voltage shown in Fig. 14.3(c) is almost distortion free. These are inverted as expected. $R_x = (R_1 || R_F)$.

Figure 14.3
Class B amplifier with shunt-shunt feedback

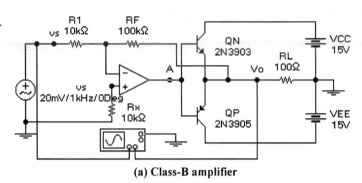

(a) Class-B amplifier

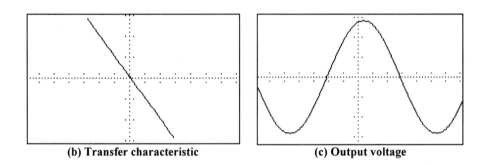

(b) Transfer characteristic (c) Output voltage

Let us apply shunt-shunt feedback to the common-emitter amplifier in Fig. 11.1. This is shown in Fig. 14.4. It should reduce both the input resistance and the output resistance, but should widen the bandwidth. The capacitor C_F and resistor R_F form the feedback network. The value of C_F should be such that it is virtually short-circuited over the frequency range of the amplifier. The feedback is taken from the output because the output is out of phase with the input voltage v_s. Since the gain of the amplifier is $A = 10^2$, the closed loop gain will be almost independent of A and it can be found from

$$A_f = \frac{V_o}{V_s} = -\frac{1}{\beta} = -\frac{R_F}{R_s} \quad (V/V) \qquad (14\text{-}10)$$

Figure 14.4
Common-emitter amplifier with shunt-shunt feedback

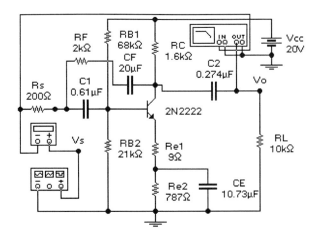

14.3   READING ASSIGNMENT

Study chapter(s) on feedback amplifiers.

14.4   ASSIGNMENT

We will design feedback amplifiers to meet either input resistance or bandwidth requirement with DC supply voltages, $V_{CC} = -V_{EE} = 20$ V.

Series-Shunt Amplifier

Design the feedback network for the amplifier shown in Fig. 14.2 to meet the following specifications:

- The input resistance with feedback must be increased by 100 times. That is, $R_{if} = 100\ R_i$.
- The voltage gain without feedback is $A = |A_v| = v_o/v_s = 10^4$.
- The voltage gain with feedback is $|A_f| = v_o/v_s = 100$.

FEEDBACK RESISTOR AND CAPACITOR

For series-shunt feedback, $R_{if} = R_i (1 + \beta A)$. The gain must decrease by a factor of 100. That is, $|A_f| = 10^4/100 = 100$. Thus, $\beta = 1/100 = 0.01$.

$$\frac{1}{\beta} = 1 + \frac{R_F}{R_{e1}} = 100$$

which, for $R_{e1} = 10\ \Omega$, gives $R_F = 1\text{k}\Omega$.

Choose a capacitor which will ensure that the C_F is virtually short-circuited at the low frequency $f_L = 1$ kHz. Choose $C_F = 20$ µF. Thus, $X_{CF} \leq 1/(2\pi \times 1\text{ kHz} \times 20\ \mu\text{F}) = 8\ \Omega$.

AMPLIFIER PARAMETERS

From Chapter 11, the input resistance without feedback is

$$R_i = R_s + R_B \| [r_\pi + R_{E1}(1 + \beta_F)] = 2613\ \Omega$$

Thus, with feedback R_{if} = 2613 x 100 = 261.3 kΩ. The output resistance without feedback is $R_o = R_{c2}$ = 1.6 kΩ. Thus, with feedback R_{of} = 1.6 kΩ/100 = 16 Ω.

$$R_{C1} = R_s + R_B \| [r_\pi + R_{e1}(1+\beta_F)] = 2613 \text{ Ω}$$

$R_{C\pi} = r_\pi \| [(R_s \| R_B) + R_{e1}(1 + \_F)] = 681.7$ Ω.
The Thevenin's equivalent resistance faced by C_μ is given by

$$R_{C\mu} = R_L \| R_C + (r_\pi \| R_B \| R_s)[1 + g_m(R_C \| R_L)] = 45.86 \text{ kΩ}$$

The low cut-off frequency without feedback is f_L = 1 kHz. From chapter 11, we get C_π = 163.5 pF, and C_μ = 2.94 pF. The high cut-off frequency f_H without feedback is given by

$$f_H = \frac{1}{2\pi(R_{C\pi}C_\pi + R_{C\mu}C_\mu)} = 485.25 \text{ kHz}$$

Shunt-Shunt Amplifier

Design the feedback network for the amplifier shown in Fig. 14.4 to meet the following specifications:

- The bandwidth with feedback must be increased by 10 times. That is, BWf = 10 BW.
- The voltage gain without feedback is A = $|A_v|$ = vo/vs = 100.
- The voltage gain with feedback is $|A_f|$ = vo/vs = 10.

FEEDBACK RESISTOR AND CAPACITOR

For shunt-shunt feedback, BW_f = BW (1 + βA). The gain must decrease by a factor of 10. Thus, 1 + βA = 10 and β = 9/A ≅ 1/Af = 0.1

$$\frac{1}{\beta} = \frac{R_F}{R_s} = 10$$

which, for R_s = 200 Ω, gives R_F = 2 kΩ.

Choose a capacitor which will ensure that C_F is virtually short-circuited at the low frequency f_L = 1 kHz. Choose C_F = 20 μF. Thus, $X_{CF} \le 1/(2\pi \times 1 \text{ kHz} \times 20 \text{ μF})$ = 8 Ω.

AMPLIFIER PARAMETERS

From Chapter 11, The input resistance without feedback is $R_i = R_\pi + R_B \| [r_\pi + R_{e1}(1+\beta_F)]$ = 2613 Ω. Thus, with feedback R_{if} = 2613/10 = 261.3 kΩ. The output resistance without feedback is $R_o = R_{c2}$ = 1.6 kΩ. Thus, with feedback R_{of} = 1.6 kΩ/10 = 160 Ω.

$$R_{C1} = R_s + R_B \| [r_\pi + R_{E1}(1+\beta_F)] = 2613 \text{ Ω}$$

$$R_{C\pi} = r_\pi \| [(R_s \| R_B) + R_{E1}(1+\beta_F)] = 68\ 1.7 \text{ Ω}$$

The Thevenin's equivalent resistance faced by C_μ is given by

$$R_{C\mu} = R_L \| R_C + \left(r_\pi \| R_B \| R_s\right)\left[1 + g_m\left(R_C \| R_L\right)\right] = 45.86 \text{ k}\Omega$$

The low cut-off (or 3-dB) frequency without feedback is $f_L = 1$ kHz. From chapter 11, we get $C_\pi = 163.5$ pF, and $C_\mu = 2.94$ pF. The high cut-off (or 3-dB) frequency f_H without feedback is given by

$$f_H = \frac{1}{2\pi\left(R_{C\pi}C_\pi + R_{C\mu}C_\mu\right)} = 646.22 \text{ kHz}$$

14.5 EWB SIMULATION

The EWB has circuit models for many types of op-amps and BJTs. First, you get the schematic of an op-amp or BJT from the list of active devices. Then, you double-click on the op-amp or BJT and the library of op-amps or BJTs will open. The library has models of many commercially available op-amps or BJTs. Also, you can change the model parameters by choosing **Edit**.

We will use EWB to simulate feedback amplifiers to verify the design specifications.

Class B Amplifier with Series-Shunt Feedback

The steps to follow are:
1. Open file FIG14_1.CA4 from the EWB file menu. Run the simulation.
2. Check that the input signal is 20 mV, 1 kHz. Otherwise, change the settings.
3. Check that the Oscilloscope has setting of DC; time base: 0.1 ms/div and Y/T; channel A: 10 mV/div; channel B: 1 V/div. Otherwise, change the settings to give a clear display.
4. Record the dead-zone on the output <u>None visible</u>. Compare the input and output waveforms and comment.

_____

_____

_____

5. Move one side of resistance R_F from node 'B' to node 'A' (that is, the base terminal of the transistors). Record the dead-zone on the output _____ (72.5 µs). What is the ratio of dead-zones, with feedback/without feedback? _____ (∞) (no detectable dead-zone for #4).

Frequency Response of Series-Shunt Feedback Amplifier

The steps to follow are:
1. Open file FIG14_2.CA4 from the EWB file menu.
2. Check that the Function Generator is set to 100 µV, 5 kHz (sine wave). Otherwise, change the settings.
3. Set the Bode plotter to magnitude, the gain on LIN (I = 0, F = 100), and the frequency on LOG (I = 10 Hz, F = 100 MHz). Run the simulation. Note that there is a 'bump' on the f_H side. Why _____
_____.
4. Use the cursor on the Bode plotter to measure f_{Lf}, f_{Hf}, and A_f. Record the values in Table 14.2.
5. To find the output resistance R_{of}, run the simulation for R_L = 10 kΩ, 10 MΩ.

$$R_o = R_L(=10\text{ k}\Omega)\left[\frac{A_f(\text{for }R_L = 10\text{ M}\Omega)}{A_f(\text{for }R_L = 10\text{ k}\Omega)} - 1\right] \quad (14\text{-}11)$$

R_o = 10 kΩ. x (96/95 - 1) = 105 Ω.

Table 14.2

	Calculated	Simulated	Practically Measured
f_{Lf} (kHz)	1	0.872	
f_{Hf} (MHz)	64.62	~24	
BW = f_{Hf} - f_{Lf} (MHz)	64.62	~24	
A_f	100	95	
R_i (kΩ)	261.36	(100µV/√2)/0.005 µA = 14.14 2	
R_{of} (Ω)	16	105	

6. Disconnect the feedback circuit, R_f and C_F. Put back to R_L = 10 kΩ. Run the simulation again.
7. Record the values in Table 14.3.

Table 14.3

	Calculated	Simulated	Practically Measured
f_L (kHz)	1	0.912	
f_H (kHz)	64.62	575	
BW = f_H - f_L (kHz)	63.62	575	
A	10^4	5.2×10^3	
R_i (Ω)	2613	$100\mu V/\sqrt{2})/0.024\ \mu A$ = 2946	
R_o (kΩ)	1.6	1.538	

If you compare the data in Table 14.3 with that in Table 14.2, you will find that the gain falls more than 10 times. Why? Explain. (Hint: Think about the loading effect of R_F).

_____

_____

_____

8. (Only for EWB version 5.0) Run the parametric sweep by varying R_F = 1 kΩ nominally (in Fig. 14.2) from 500 Ω to 2.5 kΩ with an increment of 1 kΩ. Set the AC options: start frequency: 100 Hz; end frequency: 100 MHz; sweep type: decade; number of points: 100; output node: 1. Plot the AC response of the output voltage. Discuss the effects of R_F.

_____

_____

_____

Frequency Response of Shunt-Shunt Feedback Amplifier

The steps to follow are:
1. Open file FIG14_4.CA4 from the EWB file menu.
2. Check that the Function Generator is set to 10 mV, 5 kHz (sine wave). Otherwise, change the settings.
3. Set the Bode plotter to magnitude, the gain on LIN (I = 0, F = 10), and the frequency on LOG (I = 10 Hz, F = 100 MHz). Run the simulation.
4. Use the cursor on the Bode plotter to measure f_{Lf}, f_{Hf}, and A_f. Record the values in Table 14.4.

Table 14.4

	Calculated	Simulated	Practically Measured
f_{Lf} (kHz)	1	0.197 (at 5.8 gain)	
f_{Hf} (MHz)	6.46	18.7 (at 5.8 gain)	
BW = f_{Hf} - f_{Lf} (MHz)	6.46	18.7	
A_f	10	8.2	
R_{if} (Ω)	261.3	(10μV/√2)/29.9 μA = 236.5	
R_{of} (Ω)*	160	122	

$R_o = 10\ k\Omega \times (8.3/8.2 - 1) = 122\ \Omega$

5. Disconnect the feedback circuit, R_F and C_F. Put back to $R_L = 10\ k\Omega$. Run the simulation again.
6. Record the values in Table 14.5.

Table 14.5

	Calculated	Simulated	Practically Measured
f_L (kHz)	1	1.07	
f_H (kHz)	250	709	
BW = f_H - f_L (kHz)	249	708	
A	100	86	
R_i (kΩ)	2.61	(10μV/√2)/2.508 μA = 2.82	
R_o (kΩ)*	1.6	1.63	

* $R_o = 10\ k\Omega \times (100/86 - 1) = 1.63\ k\Omega$.

If you compare the data in Table 14.4 with that in Table 14.3, you will find that the gain falls more than 10 times. Why? Explain. (Hint: Think about the loading effect of R_F).

_____

_____

_____

14.6 WRITE-UP/CONCLUSIONS

Write a brief report, summarizing the results of the design experiment and what you have learned or confirmed about feedback amplifiers.

Comment on the relationship between theory, simulation and practical circuit performance and the precautions that must be taken when using this learning approach.

_____
_____
_____
_____
_____
_____

14.7 REINFORCEMENT EXERCISES

1. The connection of the feedback circuit in Fig. 14.2 has been changed to series-series feedback. This is shown in Fig. 14.5.

The steps to follow are:

A. Open file FIG14_5.CA4 from the EWB file menu.
B. Check that the Function Generator is set to 100 µV, 5 kHz (sine wave). Otherwise, change the settings.
C. Set the Bode plotter to magnitude, the gain on LIN (I = 0, F = 100), and the frequency on LOG (I = 10 Hz, F = 100 MHz). Run the simulation.
D. Use the cursor on the Bode plotter to measure f_{Lf}, f_{Hf}, and A_f. Record the values in Table 14.6.

Table 14.6

	Calculated	Simulated	Practically Measured
f_{Lf} (kHz)		2.3	
f_{Hf} (kHz)		135	
BW = f_{Hf} - f_{Lf} (kHz)		132.7	
A_f		82.6 dB	
R_{if} (kΩ)		(100µV/√2)/0.039 µA = 1.81	
R_{of} (Ω)			

E. Calculate the feedback factor β from

$$\beta = \frac{R_{e1} R_{e3}}{R_{e1} + R_{e3} + R_F} \qquad (14\text{-}12)$$

Calculate the value of β from the simulated results in Table 14.6.
β _____

Figure 14.5
Two-stage common-emitter amplifier with series-series feedback

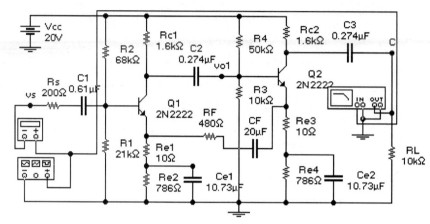

2. The connection of the feedback circuit in Fig. 14.4 has been changed to series-series feedback. This is shown in Fig. 14.6.
The steps to follow are:
A. Open file FIG15_6.CA4 from the EWB file menu.
B. Check that the Function Generator is set to 10 mV, 5 kHz (sine wave). Otherwise, change the settings.
C. Set the Bode plotter to magnitude, the gain on LIN (I = 0, F = 100), and the frequency on LOG (I = 10 Hz, F = 100 MHz). Run the simulation.
D. Use the cursor on the Bode plotter to measure f_{Lf}, f_{Hf}, and A_f. Record the values in Table 14.7.
E. Calculate the feedback factor β from

$$\beta = \frac{R_{e1}}{R_{e1} + R_F} \qquad (14\text{-}13)$$

Calculate the value of β from the simulated results in Table 14.6.
β _____

Table 14.7

	Calculated	Simulated	Practically Measured
f_{Lf} (kHz)		0.776	
f_{Hf} (MHz)		2.35	
BW = f_{Hf} - f_{Lf} (MHz)		2.35	
A_f		58	
R_{if} (kΩ)		(10 mV/$\sqrt{2}$)/0.041 μA = 172.47	
R_{of} (Ω)			

Figure 14.6
Common-emitter amplifier with shunt-series feedback

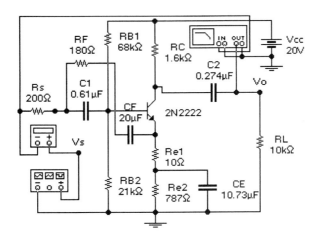

14.8 DESIGN PROBLEMS

There may be more than one solution to the following problems. Use EWB or PSPICE to verify your design. Also, build and test, if possible. Determine the voltage and current ratings of active and passive components.

1. Apply feedback and modify the design Prob. 12.1. The input resistance must be increased by 20 times, that is, $R_{if} \geq 20\, R_i$, and the output resistance be decreased by 20 times, that is, $R_{of} = R_o/20$.

2. Apply feedback and modify the design Prob. 12.2. The input resistance must be increased by 10 times, that is, $R_{if} \geq 10\, R_i$, and the output resistance be decreased by 10 times, that is, $R_{of} \leq R_o/10$.

15 DESIGN OF OSCILLATORS

15.1 LEARNING OBJECTIVES

To design oscillators to give a sinusoidal output voltage of specified frequency. We will use Electronics Workbench to verify the design and evaluate the performance of oscillators.

At the end of this lab, you will

- Be familiar with four types of oscillators and their conditions for sustained oscillations.
- Be able to analyze and design sinusoidal oscillators.

15.2 THEORY

An amplifier with negative feedback will be unstable if the magnitude of the loop-gain is greater than or equal to one, and its phase shift is $\pm 180°$. Under these conditions, the feedback becomes positive and the output of the amplifier oscillates. An oscillator is a circuit that generates a repetitive waveform of fixed amplitude at a fixed frequency without any external input signal. This characteristic is normally accomplished by employing positive feedback in amplifiers. Positive feedback provides enough feedback signal to maintain oscillations. The closed-loop voltage gain A_f with positive feedback is given by

$$A_f = \frac{V_o}{V_{in}} = \frac{A}{1 - A\beta} \tag{15-1}$$

which can be made very large by making $(1 - A\beta) = 0$. That is, an output of reasonable magnitude can be obtained with a very small-value of input signal, tending to be zero. Thus, the amplifier will be unstable under the following condition:

$$1 - A\beta = 0 \tag{15-2}$$

which gives the loop-gain as, $A\beta = 1$. This is, in polar form,

$$A\beta = 1 \angle 0° \text{ or } \angle 360° \tag{15-3}$$

This gives the design criteria for oscillation as:
- The magnitude of the loop-gain $|A\beta|$ must be unity, or slightly larger at the desired oscillation frequency.
- The total phase shift ϕ of the loop-gain must be equal to $0°$, or $360°$ at the same frequency.

Phase Shift Oscillator

A phase-shift oscillator consisting of an inverting op-amp amplifier with positive feedback is shown in Fig. 15.1. The amplifier gives a phase-shift of 180°, and the feedback circuit gives another phase-shift of 180° so that the total phase-shift around the loop is 360°.

For $R_2 = R_3 = R_4 = R$, and $C_1 = C_2 = C_3 = C$, the oscillation frequency ω_o is given by

$$\omega_o = \omega = 2\pi f_o = \frac{1}{\sqrt{6}RC} \quad (15\text{-}4)$$

The condition for sustained oscillations is given by

$$\frac{R_F}{R_1} \geq 29 \quad (15\text{-}5)$$

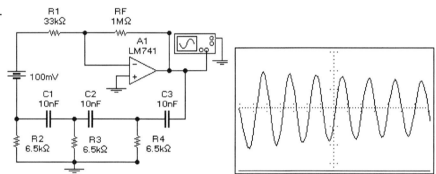

Figure 15.1 **Phase-shift oscillator**

Generally, $R_F > 29R_1$. Setting the loop-gain to unity is not a reliable method for designing an oscillator because the oscillation may not sustain the amplitude of the waveform as shown in Fig. 15.1. In practice, an oscillator is designed with a value of $|A\beta|$ that is slightly higher, say by 5%, than unity at the oscillating frequency. The greater the value of $|A\beta|$, the greater will the amplitude of the output signal and also the amount of its distortion. It is often necessary to stabilize an oscillator by introducing non-linearity with Zener diodes (or diodes) as shown in Fig. 15.2. This will ensure oscillation with predictable output voltage waveform. The small voltage V_s is used for start up. In practical oscillators, the oscillation is generally initiated by transients caused by switching of the power supply and the non-linearity of op-amp.

Quadrature Oscillator

A quadrature-oscillator, shown in Fig. 15.3, requires dual op-amps. Amplifier A_2 operates as an inverting integrator and provides a phase shift of -270° (or 90°); whereas, amplifier A_1 in combination with the feedback network operates as a non-inverting integrator and provides the remaining -90° (or 270°) to give the total phase-shift of 360° that is required to satisfy the condition of oscillation. This oscillator generates two signals (sine and cosine) which are in quadrature, that is, out of phase by 90°. The actual location of sine and cosine waveforms is arbitrary.

For $R_1 = R_2 = R_F = R$, and $C_1 = C_2 = C_F = C$, the oscillation frequency ω_o is

$$\omega_o = \omega = 2\pi f_o = \frac{1}{RC} \qquad (15\text{-}6)$$

Therefore, the overall gain A_V of amplifiers A_1 and A_2 is given by

$$A_V = \frac{1}{\beta} = \sqrt{2} = 1.4142 \qquad (15\text{-}7)$$

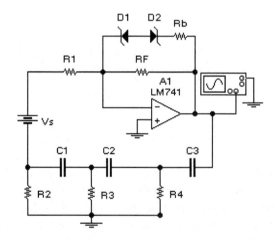

Figure 15.2
Stabilization of phase-shift oscillator

Generally, $R_F > R_1$. This oscillator can be stabilized by connecting two back-to-back Zener diodes (as shown) across one of the integrating capacitors.

Figure 15.3
Quadrature oscillator with stabilization network

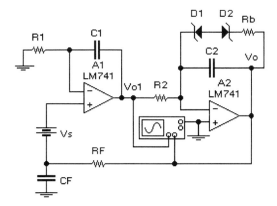

Wien-Bridge Oscillator

The *Wien-bridge oscillator* is one of the most commonly used audio frequency oscillators. It is shown in Fig. 15.4. Since the feedback signal is applied to the positive (+) terminal, the op-amp along with R_1 and R_F operates as a noninverting amplifier.

For $R_2 = R_3 = R$, and $C_2 = C_3 = C$, the oscillation frequency ω_o is given by

$$\omega_o = \omega = 2\pi f_o = \frac{1}{RC} \tag{15-8}$$

The condition for sustained oscillations is

$$\frac{R_F}{R_1} \geq 2 \tag{15-9}$$

Generally, $R_F > 2R_1$. This oscillator can be stabilized by connecting two back-to-back diodes (as shown) across the resistor, R_F.

Figure 15.4
Wien-bridge oscillator with stabilization network

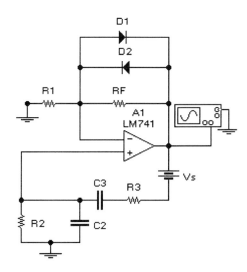

15.3 READING ASSIGNMENT

Study chapter(s) on oscillators.

15.4 ASSIGNMENT

Design Specifications

We will design the phase-shift, quadrature and Wein-bridge oscillators. The design specifications are:

- Oscillation frequency, f_o = 1 kHz.
- Use LM741 op-amps.
- DC supply voltages, V_{CC} = -V_{EE} = 20 V.

Phase-Shift Oscillator

Design the phase-shift oscillator shown in Fig. 15.2. The peak output voltage is $V_{o(peak)}$ = 5 V.

RESISTORS AND CAPACITORS

Let $R_2 = R_3 = R_4 = R$, and $C_1 = C_2 = C_3 = C$. Choose C = 10 nF. The value of R becomes

$$R = \frac{1}{2\pi\sqrt{6}f_oC} = \frac{1}{2\pi \times \sqrt{6} \times 1 \text{ kHz} \times 10 \text{ nF}} = 6.5 \text{ k}\Omega$$

Let R_1 = 10 kΩ, and R_F/R_1 = 40, which gives R_F = 40 R_1 = 400 kΩ. Choose Zener diodes of 5 V, and let R_b = 400 kΩ.

Quadrature Oscillator

Design the quadrature oscillator shown in Fig. 15.3. The peak output voltage is $V_{o(peak)}$ = 5 V.

RESISTORS AND CAPACITORS

Let $R_1 = R_2 = R$, and $C_1 = C_2 = C$. Choose C = 15 nF. The value of R becomes

$$R = \frac{1}{2\pi f_oC} = \frac{1}{2\pi \times 1 \text{ kHz} \times 15 \text{ nF}} = 10.5 \text{ k}\Omega$$

Let R_1 = 10.5 kΩ, and R_F/R_1 = $\sqrt{2}$, which gives R_F = $\sqrt{2}$ R_1 = 15 kΩ. Choose Zener diodes of 5 V, and let R_b = 10.5 kΩ.

Wien-Bridge Oscillator

Design the Wien-bridge oscillator shown in Fig. 15.4. The peak output voltage is $V_{o(peak)}$ = 0.75 V.

RESISTORS AND CAPACITORS

Let $R_2 = R_3 = R$, and $C_2 = C_3 = C$. Choose C = 10 nF. The value of R becomes

$$R = \frac{1}{2\pi f_oC} = \frac{1}{2\pi \times 1 \text{ kHz} \times 10 \text{ nF}} = 15.9 \text{ k}\Omega$$

Let R_1 = 10 kΩ, and R_F/R_1 > 2. Choose R_F = 21 kΩ. Choose diodes of 0.75 V.

15.5 EWB SIMULATION

We will use EWB to simulate the oscillators to verify the design specifications.

Phase-Shift Oscillator

The steps to follow are:

1. Open file FIG15_1.CA4 from the EWB file menu.

2. Check that the Oscilloscope has settings of AC; time base: 0.5 ms/div and Y/T; channel B: 500 mV/div. Otherwise, change the settings to give a clear display.

3. Set $R_1 = 33$ kΩ. Run the simulation by turning on the switch. Observe the output voltage carefully. Does the oscillation sustain? Yes/No: _____ Increasing/decaying/sustaining oscillation: _____

4. Change to $R_1 = 60$ kΩ by double-clicking on R_1. Run the simulation. Observe the output voltage. Does the oscillation sustain? Yes/No: _____ Increasing/decaying/sustaining oscillation: _____

5. Change to $R_1 = 25$ kΩ by double-clicking on R_1. Change the channel B: 10 V/div. Run the simulation. Observe the output voltage. Does the oscillation sustain? Yes/No: _____. Increasing/decaying/sustaining oscillation:

   _____ Sustains but not sinusaided-peaks are distorted. _____

   _____ Put scope to DC → negative peaks saturate @ -21V _____

   _____ There is a -DC offset to the oscillation V. _____

6. Comment on the effect of R_1 (that is, the ratio R_F/R_1) on the oscillation and the output voltage.

   _____

   _____

   _____

Stabilized Phase-Shift Oscillator

The steps to follow are:

1. Open file FIG15_2.CA4 from the EWB file menu.

2. Check that the Oscilloscope has settings of AC; time base: 0.2 ms/div and Y/T; channel B: 2V/div. Otherwise, change the settings to give a clear display. Run the simulation.

3. Zoom the oscilloscope display. Use the cursors to read the peak output voltage $V_{o(peak)}$, and the period of oscillation T_o ($=1/f_o$). Complete Table 15.1.

Table 15.1

	Calculated	Simulated	Practically Measured
f_o (kHz)	1	1.096	
$V_{o(peak)}$ (V)	5	4.38V	

4. If you were to change the oscillation frequency, which element(s) would you change? Explain.

_____

_____

_____

5. If you were to change the peak output voltage, which element(s) would you change? Explain.

_____

_____

_____

Quadrature Oscillator

The steps to follow are:

1. Open file FIG15_3.CA4 from the EWB file menu.

2. Check that the Oscilloscope has settings of AC; time base: 0.1 ms/div and Y/T; channel A: 2V/div; channel B: 2V/div. Otherwise, change the settings to give a clear display. Run the simulation. *If the simulation does not reach the steady-state condition, stop the simulation by clicking the stop button.*

3. Zoom the oscilloscope display. Use the cursors to read the peak voltage $V_{o1(peak)}$ at the output of first amplifier, the peak voltage $V_{o(peak)}$ at the output of second amplifier, and the period of oscillation T_o ($=1/f_o$). Complete Table 15.2.

Table 15.2

	Calculated	Simulated	Practically Measured
f_o (kHz)	1	0.911	
$V_{o1(peak)}$ (V)	5	5.8V	
$V_{o(peak)}$ (V)	5	5.7V	

4. Observe the two waveforms v_{o1} and v_o carefully. Is one a sine wave and the other a cosine wave (that is, phase-shifted by 90°)? Record the phase shift ϕ: _____ (117) °. Comment:

_____

_____

_____

5. If you were to change the oscillation frequency, which element(s) would you change? Explain.

_____

_____

_____

6. If you were to change the peak output voltage, which element(s) would you change? Explain.

_____

_____

_____

Wien-Bridge Oscillator

The steps to follow are:
1. Open file FIG15_4.CA4 from the EWB file menu.
2. Check that the Oscilloscope has settings of AC; time base: 0.1 ms/div and Y/T; channel B: 500 mV/div. Otherwise, change the settings to give a clear display. Run the simulation. *If the simulation does not reach the steady-state condition, stop the simulation by clicking the stop button.*
3. Zoom the oscilloscope display. Use the cursors to read the peak output voltage $V_{o(peak)}$, and the period of oscillation T_o (=1/f_o). Complete Table 15.3.

Table 15.3

	Calculated	Simulated	Practically Measured
f_o (kHz)	1	0.99	
$V_{o(peak)}$ (V)	0.75	0.736V	

4. If you were to change the oscillation frequency, which element(s) would you change? Explain.

_____

_____

_____

5. If you were to change the peak output voltage, which element(s) would you change? Explain.

_____

_____

_____

15.6 WRITE-UP/CONCLUSIONS

Write a brief report, summarizing the results of the design experiment and what you have learned or confirmed about the design of oscillators.

Comment on the relationship between theory, simulation and practical circuit performance and the precautions that must be taken when using this learning approach.

_____

_____

_____

_____

_____

_____

15.7 REINFORCEMENT EXERCISES

1. If positive feedback is applied to the narrow-band filter shown in Fig. 13.8, it can be operated as an oscillator. This is shown in Fig. 15.5. The feedback circuit consists of resistor R_F, and the limiting diodes D_1 and D_2. The inverting amplifier A_2 gives a phase-shift of 180° for positive feedback. The output of the diode limiter should give a square wave, which is feedback to the filter whose output is then a sinusoidal voltage at a specified frequency.

Figure 15.5
Narrow-band filter as an oscillator

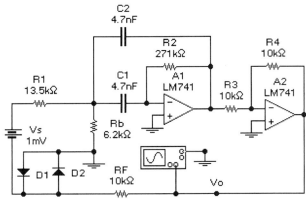

The steps to follow are:
A. Open file FIG15_5.CA4 from the EWB file menu.
B. Check that the Oscilloscope has settings of DC; time base: 0.1 ms/div and Y/T; channel B: 5 V/div. Otherwise, change the settings to give a clear display. Run the simulation.
C. Zoom the oscilloscope display. Use the cursors to read the peak output voltage $V_{o(peak)}$, and the period of oscillation T_o ($=1/f_o$). Complete Table 15.4.

Table 15.4

	Calculated	Simulated	Practically Measured
f_o (Hz)	1000	981	
$V_{o(peak)}$ (V)	10	7.78	

D. If you were to change the oscillation frequency, which element(s) would you change? Explain.

_____

_____

_____

E. If you were to change the peak output voltage, which element(s) would you change? Explain.

_____

_____

_____

15.8 DESIGN PROBLEMS

There may be more than one solution to the following problems. Use EWB or PSPICE to verify your design. Also, build and test, if possible. Determine the voltage and current ratings of active and passive components.

1. Design a phase-shift oscillator so that the oscillation frequency is f_o = 10 kHz.
2. Design an Wien-bridge oscillator so that the oscillation frequency is f_o = 5 kHz.
3. Apply feedback and modify the design Prob. 14.1 so that the amplifier oscillates at a frequency of f_o = 20 kHz.
4. Apply feedback and modify the design Prob. 14.2 so that the amplifier oscillates at a frequency of f_o = 20 kHz.